THE TERRY LECTURES

Islam, Science, and the Challenge of History

Other Volumes in the Terry Lecture Series

Available from Yale University Press

The Courage to Be Paul Tillich

Psychoanalysis and Religion Erich Fromm

Becoming Gordon W. Allport

A Common Faith John Dewey

Education at the Crossroads Jacques Maritain

Psychology and Religion Carl G. Jung

Freud and Philosophy Paul Ricoeur

Freud and the Problem of God Hans Küng

Master Control Genes in Development and Evolution Walter J. Gehring

Belief in God in an Age of Science John Polkinghorne

Israelis and the Jewish Tradition David Hartman

The Empirical Stance Bas C. van Fraassen

One World: The Ethics of Globalization Peter Singer

Exorcism and Enlightenment H. C. Erik Midelfort

Reason, Faith, and Revolution: Reflections on the God Debate Terry Eagleton

Thinking in Circles: An Essay on Ring Composition Mary Douglas

The Religion and Science Debate: Why Does It Continue? Edited by Harold W. Attridge

Natural Reflections: Human Cognition at the Nexus of Science and Religion Barbara Herrnstein Smith

Absence of Mind: The Dispelling of Inwardness from the Modern Myth of the Self Marilynne Robinson

Islam, Science, and the Challenge of History

Ahmad Dallal

Yale
UNIVERSITY PRESS
New Haven & London

Published with assistance from the foundation established in memory of Amasa Stone Mather of the Class of 1907, Yale College.

Yale University Press books may be purchased in quantity for educational, business, or promotional use. For information, please e-mail sales.press@yale.edu (U.S. office) or sales@yaleup.co.uk (U.K. office).

Designed by James J. Johnson and set in Fournier type by Tseng Information Systems, Inc.

Library of Congress Cataloging-in-Publication Data

Dallal, Ahmad S.
Islam, science, and the challenge of history / Ahmad Dallal.
p. cm. — (Terry lectures)
Includes bibliographical references and index.
ISBN 978-0-300-15911-0 (alk. paper)

ISBN 978-0-300-17771-8 (pbk)

1. Islam and science — History.
2. Science — Islamic countries — History. I. Title.
BP190.5.S3D35 2010
297.2′65 — dc22
2009047218

A catalogue record for this book is available from the British Library.

10 9 8 7 6 5 4 3 2 1

To Shezza and Millal
and Dalal

The Dwight Harrington Terry Foundation

Lectures on Religion in the Light of

Science and Philosophy

The deed of gift declares that "the object of this foundation is not the promotion of scientific investigation and discovery, but rather the assimilation and interpretation of that which has been or shall be hereafter discovered, and its application to human welfare, especially by the building of the truths of science and philosophy into the structure of a broadened and purified religion. The founder believes that such a religion will greatly stimulate intelligent effort for the improvement of human conditions and the advancement of the race in strength and excellence of character. To this end it is desired that a series of lectures be given by men eminent in their respective departments, on ethics, the history of civilization and religion, biblical research, all sciences and branches of knowledge which have an important bearing on the subject, all the great laws of nature, especially of evolution . . . also such interpretations of literature and sociology as are in accord with the spirit of this foundation, to the end that the Christian spirit may be nurtured in the fullest light of the world's knowledge and that mankind may be helped to attain its highest possible welfare and happiness upon this earth." The present work constitutes the latest volume published on this foundation.

Contents

Preface xi

Chapter 1. Beginnings and Beyond 1

Chapter 2. Science and Philosophy 54

Chapter 3. Science and Religion 110

Chapter 4. In the Shadow of Modernity 149

Notes 177

Index 227

Preface

This book is an attempt to situate the Islamic culture of science in relation to social and cultural trends in Muslim societies. Needless to say, a task of this nature requires broad generalizations about numerous disciplines of learning produced in a vast region over the span of many centuries. In the course of generalizing about macrohistorical trends, I often gloss over the historical specificities of books, authors, and disciplines. Yet, while many more detailed studies are still needed before a comprehensive history that accounts for such specificities can be attempted, I think there is a rationale for, and value in, the types of generalization offered in this book. Numerous specialized studies of science in Muslim societies have been produced in the past three decades, providing a nuanced understanding of important trends in the history of a range of scientific disciplines. The current level of knowledge deriving from these studies warrants, in my view, venturing further generalizations. While often tentative, these generalizations will, I hope, provide a lens through which concrete cases can be further scrutinized and elucidated.

The book is based on the Terry Lectures that I gave at Yale University in February 2008. I am grateful for the invitation to give these lectures and for the feedback provided by the members of the Dwight H. Terry Lectureship Committee. I am also grateful for the feedback of the audience attending these lectures, and especially for the valuable comments and suggestions by Professors Dimitri Gutas and Frank Griffel. Needless to say, I am solely responsible for the shortcomings of this book.

I am also thankful to Jean Thomson Black and the editorial team at Yale University Press for their help and support.

I had the honor of being named a Carnegie Scholar in 2007, and the research for this publication was made possible by a grant from the Carnegie Corporation of New York. The corporation is not responsible for the statements and views expressed in the book.

I wrote most of this book in 2007–8 in Casablanca, where I enjoyed the loving support of my wife, Dalal, my daughter, Shezza, and her brother, Millal. The warmth of their presence and the pleasure of their company made the year a joyful experience. In gratitude I dedicate this book to them.

Islam, Science, and the Challenge of History

Chapter 1

Beginnings and Beyond

Muslims are enjoined to face Mecca during their five daily prayers, just as all mosques are supposed to be oriented toward the Ka'ba in Mecca, in what is known as the direction of the *qibla*.[1] Before mathematical methods became available, Muslims determined the direction of the qibla based on the practices of the early Companions of the Prophet Muḥammad and their successors. They also made use of traditions of folk astronomy and of the astronomical alignment of the Ka'ba itself. These methods provided reasonable approximations in locations close to Mecca but were quite inaccurate in faraway places like North Africa and Iran.

With the emergence of mathematical sciences, new methods of considerable sophistication were devised to compute the qibla for any locality on the basis of its geographical coordinates and Mecca's. In the most accurate solution, the problem is transferred to the celestial sphere, where the position of the zenith of Mecca relative to the zenith of the locality is determined. The zenith is determined by extending a line from the center of the earth through the position of an observer at the locality and perpendicular to the observer's horizon.[2] The direction of the qibla is then calculated as the azimuth of the zenith of Mecca on the local horizon, where the azimuth is the arc of the horizon between two great circles of the celestial sphere whose planes are perpendicular to the horizon: the vertical circle (meridian) that passes through the two poles and the observer's zenith, and the vertical circle that passes through the zenith of Mecca.[3] Most astronomical handbooks contained chapters on finding the direction of the qibla by one or more approximative or accurate methods, separate treatises were composed on the subject, and tables were published displaying the qibla direction as a function of terrestrial longitude and latitude, thereby providing the results of complex mathematical computations.

In many works on "astronomy in the service of Islam," David King has noted that jurists and scientists often proposed different solutions to the same problem, but jurists criticized mathematical astronomy only when it was used in astrology and, with occasional exceptions, did not criticize exact mathematical methods that differed from their own methods, nor did scientists often speak against the simplified methods of the religious scholars.[4] In other words, alternative methods of radically different provenance, some relying on religious tradition and others on mathematical astronomy, were usually tolerated. Methods used to determine the direction of the qibla, however, were a major exception to this general rule.

Beyond the Hijaz (by the Red Sea), Syria, and Iraq, which were near enough to Mecca so that pre-mathematical methods of computing the direction of the qibla provided fairly accurate results, many of the mosques built in the early period of Islamic expansion were misaligned. With increased knowledge of mathematical astronomy, this flaw was recognized, and although some of the misaligned mosques retained their orientation, others were rebuilt to face in the correct direction. This presented a serious problem—namely, the possibility of tearing down mosques built on the authority of the Companions of the Prophet on the basis of the findings of mathematical astronomy. More generally, the question raised was whether mathematical knowledge should take precedence over religious authority in a matter where, admittedly, the realms of science and religion overlapped.

The problem of the direction of the *qiblat ahl Fās,* the qibla of Fes, was debated for several centuries.[5] I do not intend to treat the problem in any detail, but I would like to use it as a way of introducing the epistemological question that I shall be addressing in this book: the relative authority of religious knowledge and scientific knowledge.[6] The sources examined for this overview date to the fourteenth century through the early eighteenth century.[7] The earliest source also provides detailed accounts of and long quotations from astronomers and religious scholars from the twelfth and thirteenth centuries.

One twelfth-century scholar maintained that those who assigned one qibla direction to all North African countries were wrong because this region is vast, with considerable changes in the altitudes and longitudes of its different parts. To corroborate his claim, this author referred to the Companions' redirection of the qibla of Fuṣṭāṭ (later Cairo).[8]

The fourteenth-century scholar Al-Maṣmūdī recognized a number of difficulties associated with finding the qibla in North Africa, in part, he says, because the religious scholars who wrote the authoritative legal works used in North Africa did not mention ways of finding the direction of the qibla using the stars and the risings and settings of the sun. A second difficulty arose from the drastic differences in the orientations of mosques; in the same city some mosques were directed to the east and others to the south. Those who directed their mosques to the south relied on a *ḥadīth* of the prophet that says, "Between the East and the West is a qibla," and took this to be a general ḥadīth, although, Al-Maṣmūdī adds, most religious scholars consider this ḥadīth to be relevant only to Medina and similar regions, such as Syria. Al-Maṣmūdī quotes Imām Mālik (d. 796), whose legal school was the dominant school in North Africa, to confirm that the ḥadīth does not have general applicability.[9]

Interestingly, Al-Maṣmūdī, who maintains that the traditional qibla is wrong, adds that he bases this assertion on the sayings of religious scholars, because the law (*shar*ʿ) was not founded on mathematics (geometry) and because only a small number of people are competent in geometry.[10] He goes on to refer to a "valuable" work on the qibla of the Maghrib (northwest Africa) that only a few people can understand because it is based on geometry. He also says that he consulted many of his colleagues who know how to extract the direction of the qibla using the astrolabe or mathematical computations, but their answers "were not accessible to the understandings of people like us."[11]

To Al-Maṣmūdī, the problem is not that mathematical computations are wrong; in fact, they are not. Those in North Africa who rely on a literalist reading of the ḥadīth and face the south in their prayers are wrong. The problem, however, is that the mathematical methods are often not accessible to the masses.[12] What he wishes to provide, therefore, are methods for determining the qibla in the Maghrib that are not only traditional-sounding and accessible to ordinary people but also compatible with mathematical findings.

One of the issues of concern for Al-Maṣmūdī and other participants in this debate was how to redirect the qibla without risking social conflict. In most cases, the deviation in the orientation of the qibla was small and did not call for the drastic measure of tearing down the mosque. In such cases, praying at a slight inclination with respect to the original mosque niche would solve the problem. If, however, the difference was substantial, then a Muslim should seek verifiable evidence and consult those who could provide it. Unless there was fear of civil strife, the misaligned mosque ought to be rebuilt in the right direction.[13]

Another important work on the subject of qibla orientation was written by Al-Tājūrī, a sixteenth-century religious scholar and timekeeper.[14] Al-Tājūrī's book includes a question soliciting the fatwas (legal rulings) of the scholars of Cairo and Egypt about the mosques of Fes that were directed toward the south, including the city's famous Qarawiyyīn mosque, the most important mosque-school complex in Morocco.[15] Al-Tājūrī maintained that these mosques were not directed toward the legal qibla (*al-qibla al-shar'iyya*), which is the eastward direction of Mecca. In addition to wrongly interpreting the above-mentioned ḥadīth, those who defended the false qibla invoked the precedent or practice of the early generations of Muslims who built the first mosques in Morocco in the presence, and with the consent, of religious scholars.[16] To be sure, these early mosques were built before the rise of a scientific culture in the Muslim world, but the religious authority of their builders was upheld even after the rise of science. Given the gravity of Al-Tājūrī's challenge, it is not surprising that he wished to muster religious support for his position. Besides soliciting the support of the scholars of Egypt, he used religious language to refer to the correct qibla, calling it the qibla of the Companions. Despite this veneer of religiosity, the real question he tackled was whether the qibla was to be determined on the basis of religious precedence or mathematical astronomy.

Al-Tājūrī was criticized and defended by several Fāsī scholars, but, for our purposes, the most interesting defense of his views was by a late seventeenth-century scholar, Al-'Arabī Ibn

'Abd al-Salām al-Fāsī, who was responding to a critique of Al-Tājūrī by a fellow Fāsī scholar. Al-Fāsī refers to a distinction made by some scholars between the *jiha* of the qibla, or its general direction, and the *samt* of the qibla, or the precise azimuth of the zenith of Mecca.[17] People who make this distinction, he says, suggest that the only requirement of the law is that Muslims face the general direction of the qibla without requiring knowledge of its exact mathematical coordinates, which would involve knowledge of the science of geometry. These scholars argue that since knowledge of geometry is not a legal obligation, no other legal obligations can be contingent on it (fols. 2, 6). In response to this rather compromising view, Al-Fāsī insists that the meanings of *jiha* and *samt* are the same, and that geometry is not different from any other commonly used skill, such as those used in construction and commerce, "because each craft (*ṣanā'i'*) that involves precision and measurement partakes in geometry. In fact, it is even possible to chastise a person capable of finding the exact direction of Mecca who leaves that and instead imitates (*yuqallid*) [the direction of] the niche (*miḥrāb*)" that was erected in the interest of people who have no knowledge of the ways of finding the direction of the qibla (fol. 6). After noting that many mosques are equipped with such astronomical instruments as the astrolabe and the Zarqālī plate, which are used for finding the direction of the qibla and for determining the times of prayer, and that numerous scholars have composed treatises on these instruments without ever being criticized for doing so, Al-Fāsī distinguishes between two senses of the term *jiha* (direction): as an objective in itself (*maqṣad*) or as a means (*wasīla*) for finding the direction (fol. 13).[18] The ultimate objective, he says, is to find the mathematical coordinates of the direction of the Ka'ba; the second sense of the term refers to approximations similar to the one indicated in the ḥadīth of the Prophet. Furthermore, "the exercise of independent legal reasoning (*ijtihād*) in matters related to the qibla is valid only through use of proofs that are suitable for finding this direction (*al-adilla al-munāsiba*) and not through guess and conjecture" (fol. 14). Al-Fāsī then refers to a legal opinion attributed to Imām Mālik, whose doctrine is that of the official legal school of North Africa: if the orientation of a mosque is based on ijtihād, then rebuilding it is not required in case of an error. In response, Al-Fāsī maintains that this is true if the ijtihād is based on proofs derived from astronomy or the use of astronomical tables but that there is no credence in an ijtihād that is not based on proofs (fol. 14).

To hold all Muslims responsible for praying in the correct direction by using mathematical knowledge that only a few can attain is contrary to the spirit of Islamic law, say some of Al-Fāsī's opponents. Al-Fāsī responds:

> Each craft has its masters, and nothing comes easy; learning [how to find] the direction of the qibla is similar to learning other sciences; in fact, it might even be easier than learning more elaborate texts, and it is attainable in a short period. In large cities, it is illegal for someone who does not know [how to find] the direction of the qibla to build a mosque, unless he is accompanied by masters of the craft who know the proofs of the qibla (*adillat al-qibla*). It is permissible to erect only [a mosque oriented in] the direction of Mecca, and someone who does not know the proofs of the qibla should not exercise his ijtihād even if he happens to be a jurist, *because the most a jurist can know in his capacity as a jurist is that it is obligatory to face the qibla and that it is obligatory for a non-mujtahid to imitate a mujtahid in this matter—that is, to imitate one who knows the suitable proofs for it.* (fols. 21–22; my emphasis)

Elsewhere, Al-Fāsī develops his point about the mujtahids, the scholars who exercise independent legal reasoning (ijtihād):

> The real mujtahids in the matter of the qibla, using proofs suitable for it (*al-mujtahidīn fī al-qibla bi-adillatihā al-mansubā ʿalyhā*), are the astronomers, not the jurists. Because . . . the prerequisite for this [ijtihād] is knowledge derived from the sciences of mathematics and mathematical astronomy (*hayʾat al-falak*), from timekeeping, and from the positions of the planets and the computation of directions, and all of these are outside the domain of legal science. . . . Furthermore, the ijtihād of astronomers in the matter of the qibla is not similar to the ijtihād of jurists in applied law, because there is only one correct outcome for the ijtihād in the qibla, whereas for jurists each mujtahid is correct in applied law. . . . This is why in the matter of the qibla the astronomers are given precedence over the jurists, because each craft has its masters, and the masters of the craft of [finding the direction] of the qibla are the astronomers. (fol. 32)

In the following fifty pages of his treatise, Al-Fāsī quotes and comments on numerous legal rulings and questions, only to reiterate that "the ijtihād of the Companions [of the Prophet] is certain, that [the ijtihād] of jurists is probable and uncertain (*iḥtimālī dhannī*), and that [the ijtihād] of the astronomers in the [matter of the] qibla is scientific and technical (*ʿilmī ṣināʿī*), which is equivalent to certainty because they [the astronomers] and not the jurists know the proofs of the qibla. For these reasons, astronomers take precedence over jurists" (fol. 81).

The epistemological questions raised in these texts reflect widespread discussions that took place in many fields over a very long period of time. Owing to the practical nature of the qibla debate, the conceptual categories under discussion may at times seem vague, but these issues were articulated with much more precision in other theoretical debates. The significance of the qibla debate is that precise epistemological discussions filtered down to the sensitive matter of prayer and raised, in no uncertain terms, the question of intellectual authority within, as it were, the most sacred space of Islam. Clearly, this was not an academic debate relegated to the margins of Islamic culture but a debate constitutive of the culture and, as I will argue, one of its characteristic features. (As noted earlier, my brief overview does not exhaust this rich and complex subject. Rather, the purpose of this introduction is to point to the epistemological questions addressed in this book.)

The primary questions here are, What cultural forces provided the conditions that made such epistemological debates possible, and what was the context in which these debates emerged and were sustained? I will base my analysis on texts. Determining the significance of scientific texts, and by extension of scientific knowledge, is contingent on situating these texts within the nexus of three systems of knowledge: religion, philosophy, and science. Any understanding of the cultural significance of Islamic scientific thought requires an evaluation of the relationship between science and philosophy and between science and religion. What is characteristic about the Islamic scientific tradition owes much to the peculiar relationship that was negotiated between these three systems of knowledge. These relationships will be addressed in the second and third chapters. But first I will explore key features of the historical context of the Arabo-Islamic culture of science.

Arabo-Islamic Scientific Culture in History

The history of the Arabo-Islamic sciences is part of a larger, nonlinear history that relates to earlier scientific traditions and that exerted significant influences on later traditions. However, the Arabo-Islamic sciences are not reducible to what came before, nor is their significance simply on account of what happened after. Genealogical and teleological approaches to the history of Islamic science are legitimate lines of inquiry, but I am more interested in examining Islamic scientific culture as a historical occurrence whose singularity derives from the peculiar ways it relates to other cultural forces in Muslim societies.

To be sure, a historical account of the Islamic sciences may choose to ignore the fate of the scientific tradition and to focus on the life of the tradition within its own historical context, but no account of this tradition can have value unless it addresses the question of beginnings, that is, the conditions under which an Arabo-Islamic scientific culture emerged.

Determining the beginning or beginnings is itself partly a function of the epistemological assumptions of the historian. Before the coming of Islam in the seventh century, and for over a century after its rise, Arabs had no science. Without exception, historians of Islamic science have rightly identified the "translation movement," the bulk of which took place in the course of the ninth century, as the most important factor in the emergence of an Islamic scientific culture. This translation movement provided the knowledge base of the emergent sciences. But while this explains part of the picture, and admittedly one of its most important parts, it does not provide a full explanation of the beginnings. To start with, what are the sociopolitical conditions and the cultural aptitudes that triggered interest in translation and science in the first place? Second, what were the cultural conditions that enabled a significant community of interest to know how to translate complex scientific texts, to develop the technical terminology needed for the transfer of scientific knowledge between two languages, to understand scientific texts once they were translated, and to constructively engage the knowledge derived from them? Seen in this light, translation is not a mechanical process but part of a complex historical process that is not reducible to the transfer of external knowledge; rather, it involves forces intrinsic to the receiving culture—most important, the epistemological conditions internal to Islamic culture at the time of the translations. And finally, the lives of the sciences enabled by the translations are related but not identical to the life of the translation movement itself. The evolving interaction of the imported sciences with complex cultural forces, including the knowledge base assembled at the moment of translation, resulted in a continuous restructuring of the sciences as integral components of an evolving Islamic culture. The moment of translation, important as it was, is only an aspect of and one factor in the subsequent development of the Arabo-Islamic sciences—hence, "Beginnings and Beyond" as the title of this chapter.[19]

The historical significance of scientific ideas and practices is shaped by multiple historical factors that an account of origins cannot possibly exhaust. Besides asking about origins, it is important to ask if and how individual scientific ideas acquired specific meanings in their new Islamic cultural context and what departures and innovations were occasioned by giving these ideas a new home. The focus here is not on precedents, despite the importance of this subject in the history of science, but on the way new ideas were received in their own cultural context, how they enabled new practices and affected further developments in their own and other disciplines, and what kinds of thinking and practice were engendered by this cumulative process.[20] This approach avoids assigning exaggerated weight to any single scientific discovery and avoids dismissing it as a happy guess, as is often done by historians of Islamic science.[21]

Since the task I am setting involves broad generalizations about a large subject, some cautionary remarks are in order. The scope of Islamic scientific activities is vast. Science in medieval Muslim societies was practiced on a scale unprecedented in earlier or even contemporary human history. In urban centers from the Atlantic to the borders of China, thousands of scientists pursued careers in many diverse scientific disciplines. Until the rise of modern science, no other civilization engaged as many scientists, produced as many scientific books, or provided as varied and sustained support for scientific activity. So one obvious question for the historian is, Where—in which regions, during which periods, and in which genres—do we look to extract evidence that can be generalized? Both scientific texts and texts about science provide the basis of the narrative that I will construct in this chapter. Building on this basis is justifiable, I think, because texts provide the best way of identifying the generic characteristics of science as a distinct kind of cultural activity, one that is at once dependent on but not reducible to other forms of cultural activity. To put it differently, the historically constituted epistemological requirements of science are best described in texts.

But even if we restrict ourselves to written products and leave aside the countless material artifacts and technological achievements, the scale remains overwhelming. This difficulty is compounded by the unavailability of the majority of Arabo-Islamic scientific manuscripts; most remain unstudied and are often even not catalogued. Enough specialized studies have been produced in the past few decades, however, to warrant the tentative generalizations proposed here.

The specialized studies of Edward Kennedy and his students on the exact sciences in Islam opened new horizons in the field after generations of nonspecialized commentaries on Islamic science by general historians who had little knowledge of the sciences.[22] Based on these and other specialized works, including his own studies, George Saliba has provided what to date is the most comprehensive attempt to provide a synthesis of the history of the reform tradition in Islamic theoretical astronomy.[23] David King has produced several important studies on practical astronomy in which he thoroughly surveys the field; he covers such topics as instruments, timekeeping, and astronomical computation.[24] The wide-ranging research of Roshdi Rashed has been instrumental in advancing our understanding of the disciplines of Arabic mathematics.[25] In medicine and the life sciences, the bibliographical overview of M. Ullmann, along with several interpretive essays by Michael Dols, Lawrence Conrad, and others, has encouraged recent attempts to provide synthesized historical overviews.[26] These and many other detailed studies have enriched our knowledge of the scientific legacies of the medieval Muslim world.[27] In addition, a number of comprehensive bibliographical works of Arabic science have established the scale of the scientific activity and its historical span. The works of Fuat Sezgin and Gerhard Endress, in addition to Ullmann's bibliographical essays, are indispensable tools in this regard.[28] The works I draw on most, however, are two interpretive essays published in 1998 whose authors have attempted to address in systematic ways the particular social and historical roots of the translation movement and the general social factors that contributed to the rise and development of scientific thought.[29]

The books published by Dimiti Gutas and George Saliba provide critical assessments of a problematic tradition of scholarship on the rise of Islamic science.[30] After identifying the major gaps in the Orientalist narratives proposed (and often imagined) by earlier historians, the authors construct alternative narratives, based on a meticulous examination of the extant historical record.

There are, to be sure, significant differences between the two scholars. One main difference is that Saliba situates the main impetus for the rise of the Arabo-Islamic sciences in the Arabization of the administrative apparatus during the mid-eighth century, that is, during the reign of the Umayyad caliph ʿAbd al-Malik (d. 705), whereas Gutas argues that the impetus derived mainly from the imperial ideology adopted by the early ʿAbbāsid caliphs, especially Al-Manṣūr (d. 775), which created an atmosphere conducive to the development of science and generated material support. This material support, according to Saliba, catered to social needs but derived mainly from the decision of the Umayyad ruler to dedicate huge resources for restructuring the state apparatus.[31] Another significant difference is that Gutas identifies interest in political astrology as an aspect of the ideological outlook of the ʿAbbāsids, whereas Saliba argues that astrology played a role in the development, but not the rise, of science.[32]

Despite these differences, the two scholars take a common approach to the question of beginnings, and the primary problematics addressed in both studies are quite similar, as are the types of questions asked. Above all, both scholars treat the translation movement and scientific activity broadly, as complex phenomena that do not lend themselves to single-track and static explanations.[33] Scientific activities, including translation, received wide support from various social classes and thus cannot be explained, as traditional accounts often suggest, in terms of the interests of any particular group or the whims and proclivities of any particular ruler. Both studies make a compelling case for situating the translation movement in the context of an emerging scientific tradition in the growing Muslim urban centers, whose culmination was Baghdad. They see translation as an aspect of this emerging scientific culture and not its mechanical cause. Both studies also stress the important role that Syriac speakers who translated from Greek to Arabic played in the translation movement because of their multilingual skills, and not because they were the preservers of a distinct Syriac scientific tradition. Both studies undermine the thesis that Greek knowledge was preserved in Syriac pockets of learning from which science reemerged in the ninth century.[34]

In what is perhaps their main contribution, these two authors recognize a major gap in earlier scholarship on Islamic science: failure to identify a living scientific tradition within Islamic cultural space that would explain the seemingly sudden appearance of a thriving scientific culture in the early decades of the ninth century. For translations to be understood and to have an impact, there must have existed a scientific culture, what I call a knowledge base, on which further knowledge could be grafted.[35] Saliba argues that during the Arabization of the administrative apparatus, a group of people acquired foundational knowledge of the basics of various sciences.[36] Gutas, on the other hand, argues that what he calls international scholars existed in the seventh and eight centuries; they were multilingual (Greek, Arabic, Syriac, Pahlavi), scientifically competent scholars working in a region now unified under Islamic rule and representing a scientific tradition alive before the period of translation. These international scholars practiced their profession "in whatever environment offered the best support . . . thereby transmitting much scientific knowledge without translation"; and when the 'Abbāsid imperial ideology provided support for increased scientific activities, a critical mass of specialists came from the ranks of these international scholars.[37]

Whether through these international scholars or through the skilled professionals involved in the Arabization of the Umayyad administration, scientific activity in the Muslim world was ready to make the leap toward a more systematic and homogenous scientific and philosophical enterprise by the beginning of the ninth century. At that time, practical social and political needs, coupled with theoretical and scholarly needs, gave rise to and nurtured the systematic translation movement that had a great impact on the subsequent development of a scientific culture in the Muslim world. As the quoted works clearly illustrate, the conditions for the rise of a scientific culture must include not only social and political factors but the factors that made interest in the sciences and subsequent scientific activities possible. To the list of factors contributing to the formation of this prerequisite knowledge base should be added the linguistic activities of early Muslim society.

Historical sources preserve ample information about the methods used by ninth-century translators. Their efforts led to the creation of a highly precise scientific terminology, turned Arabic into a universal language of science through which several scientific traditions were fused, and enabled a level of scientific exchange unprecedented in earlier civilizations. Before these ninth-century translations, the earliest scholarly productions among Muslims were of a linguistic nature. Of particular relevance to the later development of science were the extensive compilations by Arabic philologists and lexicographers. The specialized lexicons that were produced represent a large-scale attempt at collecting and classifying the knowledge of the Arabs. These attempts were not always "scientific," and later, more systematic achievements eclipsed them. Nonetheless, these encyclopedic efforts provided a linguistic foundation for the development of various disciplines. The foundational philological work done by the early lexicographers was a first step in organizing knowledge and producing a scientific culture.[38]

In certain cases, momentum for scientific research derived from interest in the language sciences themselves. Combinatorial analysis was one such research area; it developed in connection with not just algebraic research but also linguistics. To compile an exhaustive Arabic lexicon, Al-Khalīl ibn Aḥmad al-Farāhīdī (d. 786), one of the earliest Arab lexicographers, enumerated, for all the letters of Arabic alphabet, all the possible combinations of words with roots of three to five letters. He first restructured the words of the lexicon on formal grounds, thereby exhausting the domain of "possible language." Defining "real language" as the "vocally actualized part of the possible language," Al-Khalīl proceeded to eliminate those words that were not actually used. By providing a theoretical solution to a practical linguistic problem, Al-Khalīl rationalized the empirical practice of lexicography. More important, his linguistic activity provided an incentive for research in the field of combinatorial analysis.[39]

Social and Institutional Contexts

The lexicography example suggests that a primary task of the historian of Islamic science is to go beyond identifying the initial conditions that might give rise to a scientific culture to identifying the historically specific characteristics of the kinds of knowledge occasioned by these conditions. Given the expansive growth of scientific activity following the translation period, the historian also needs to explore how science was able to sustain its cumulative growth in different and continuously changing historical contexts independently of any single factor or specific constellation of factors that contributed to its rise in the first place.[40] The unfolding epistemological and cultural contexts of the practice of science in Muslim societies will be explored in the next two chapters. First, however, let me briefly outline some of the social and institutional contexts for the ongoing development of science.

One of the most commonly repeated assertions in general histories of science is that the rational sciences in Muslim societies were marginalized because of the lack of institutional support.[41] The primary evidence for marginalization is adduced from the alleged exclusion of the rational sciences from the curricula of formal institutions of learning. In a classic work on the rise of colleges, George Makdisi contends that the quintessential institutions of learning in Muslim societies, madrasas and their antecedent and cognate institutions, were exclusively devoted to the study of the legal sciences and other ancillary religious and philological disciplines and had no room for the rational sciences.[42] Makdisi recognizes the profound influence of Greek works on the development of Islamic thought and education.[43] However, he maintains that the struggle for the rational sciences was "uphill," and that the "godless sciences of the ancients" were effectively excluded from the curricula of the formal institutions of learning. As a result, the "'foreign sciences' had to be pursued privately," and "they were not subsidized in the same manner as the Islamic sciences and its ancillaries," although "there was nothing to stop the subsidized student from studying the foreign sciences unaided, or learning in secret from masters teaching in the privacy of their homes, or in the *waqf* [endowed] institutions, outside of the regular curriculum."[44]

There has been serious criticism of Makdisi, as well as attempts to redeem his general thesis about Islamic institutions of learning. The gist of the criticism is that even religious education was unsystematic, personal, and informal and that teaching institutions played no significant role in Islamic education. Based on the relative freedom of Muslim scholars in their choice of which disciplines and books to teach, critics of Makdisi argue that the structure of Muslim education has to be understood in terms of networks of individuals rather than institutions. This argument undermines the notion of formal institutional education in the religious sciences and reinforces the informality and marginality thesis with regard to the rational sciences.[45]

Scholars of Islamic education mostly agree on the marginality of the sciences. A notable exception is Sonja Brentjes, who has surveyed social and cultural spaces for the practice of the rational sciences in Muslim societies.[46] She argues that education in the exact sciences was a stable aspect of education in Muslim societies, just as education in the religious sciences was. Furthermore, scientific education took place within recognizable networks whose structures paralleled those of religious education. This observation is borne out by information culled from the biographical dictionaries and by the abundant use in these sources of parallel discourses to describe education in the rational as well as the traditional sciences. In these discourses on education and knowledge, the rational sciences are always present and treated with respect, and they are often presented in alliance with the religious sciences. In particular, Brentjes notes the alliances and professional affiliations between several religious and rational sciences after the thirteenth century.[47]

In fact, the evidence for the presence of the rational sciences as a constitutive element in the educational landscape of classical Muslim societies is simply overwhelming. The testimony for their constant presence comes mainly from the actual combination of religious and scientific scholarship in the persons of numerous scholars, especially after the twelfth century. In different regions and periods, countless scholars produced advanced scholarship in such fields as ḥadīth and medicine,[48] Qur'ānic exegesis and astronomy,[49] and law and philosophy, not to mention theology, grammar, logic, and many other areas of study.[50] Scholars often alternated teaching religious and scientific disciplines in the same place.[51] Moreover, the rational sciences were always included in the many classifications of science that proliferated in Islamic scholarship.[52]

The shifting professional alliances between religious and scientific disciplines in the actual practice of individual intellectuals in the classical Islamic world, and the shifting ties between disciplines, provided social prestige and respectability to the rational sciences. Similarly, the creation of subfields within the scientific disciplines, subfields that addressed specific problems related to Muslim ritual and that were integrated into religious education, bestowed on these scientific disciplines a status similar to that enjoyed by the religious sciences. Two notable examples of such disciplines are the science of *farā'id,* which dealt with legal inheritance computations according to Islamic law and was a subfield of algebra, and *'ilm al-mīqāt,* or timekeeping, which dealt with computations of times of prayer and computations for calendars and was a subfield of practical or applied astronomy.[53]

Further evidence for the respectability of the rational sciences is found in the abundant side-by-side appearances in biographical dictionaries of scholars of the rational and religious sciences. In countless entries in biographical dictionaries dedicated to Islamic law, the study of ḥadīth, or other religious disciplines command of the rational sciences is celebrated as a positive trait.[54] A significant number of entries also mention that their subjects studied or taught one or more of the secular sciences and wrote books in these fields, and similar expressions are used to praise a scholar's knowledge in both rational and religious sciences.

The examples cited so far are from biographical dictionaries dedicated to religious scholars. Equally important is the emergence of specialized kinds of biographical dictionaries dedicated to physicians (*ṭabaqāt al-Aṭṭibā'*) or practitioners of the rational sciences (*ṭabaqāt al-ḥukamā'*). As with other kinds of biographical dictionaries, the specialized biographies reflect a community's awareness of its own professional and intellectual identity. The dictionaries themselves amount to a community's attempts to legitimize its own knowledge and present it as a worthy achievement within the larger Muslim community, thereby situating a particular professional practice within the mainstream of Islamic culture.[55] Thus, for example, one biographical dictionary of physicians presents medical authorities, including Greek physicians, as virtuous sages, an ideal type sanctioned by Islam.[56]

Through these and other means, the pursuit of science was conceptualized as an act with social and epistemological value in its own right, and various structures were instituted to promote and support scientific activity. As already noted, science was taught in a variety of venues, ranging from private homes to endowed buildings. It was likewise practiced in various locales, ranging from private homes to formal institutions. Medicine, for example, was taught privately in the homes of individual physicians, but it was also taught systematically in endowed hospitals, and beginning in the twelfth or thirteenth century, it was also taught in endowed schools established exclusively for the teaching of science.[57]

The prestige of the medical profession depended on mechanisms of social and professional integration, most notably through hospitals. The hospital was one of the greatest institutional achievements of medieval Islamic societies. Between the ninth and the tenth centuries, five hospitals were built in Baghdad alone, and several other hospitals were built in a number of other regional centers. The most famous is the ʿAḍuḍī hospital, established under Buwayhid rule in 982. After the tenth century, the numbers of hospitals increased significantly; such prominent institutions as the Nūrī hospital of Damascus (twelfth century) and the Manṣūrī hospital of Cairo (thirteenth century) were built, along with others in Qayrawān, Mecca, Medina, Rayy, and many more places. Many of these hospitals were divided into sections: men and women were treated in separate halls; special areas were reserved for the treatment of contagious diseases; separate areas were designated for surgical cases and for the mentally ill. The hospitals also had living quarters for the physicians in attendance, as well as for other members of the service teams. Some hospitals had their own pharmacies; and some, libraries that could be used for medical instruction. Clinical training and bedside instruction were often provided. A chief administrator, who usually was not a physician, ran the hospital, while a chief of staff, who was also the head physician, ran the medical side. Many of these hospitals had huge operating budgets, which were usually derived from the revenues of waqf properties dedicated for the hospitals at the time of their establishment. The revenues were spent on the maintenance of the premises and the salaries of the staff, as well as the cost of treatment, which was provided free of charge.

Unlike medicine, mathematics and astronomy were partly integrated into the curricula of religious schools, through the disciplines of farā'iḍ (algebra of inheritance) and ilm al-mīqāt (timekeeping).[58] And while we have some evidence for the teaching of theoretical astronomy in religious schools after the thirteenth century, we are much more informed about the institutional framework for the practice of the science of astronomy.[59] One such institution where it was practiced was the observatory.[60]

The earliest planned and programmed astronomical observations were produced during the last years of the reign of Al-Ma'mūn (r. 813–33), at the outset of the translation movement. Under al-Ma'mūn, a program of astronomical observation was organized in Baghdad (Shammāsiyya) and Damascus (Mount

Qāsiyūn). Like any organized research project, this one endowed astronomical activity in the Islamic world with formal prestige. It also set a precedent for future support of scientific activity by other rulers and established patronage as one of the modes of supporting scientific activity. The professed purpose of the program was to verify and correct the Ptolemaic observations for the sun and the moon by comparing the results derived by calculation, based on the Ptolemaic models, with actual observations conducted in Baghdad and Damascus some seven hundred years after Ptolemy.[61] The results were compiled in *Al-Zīj al-Mumtaḥan* (The Verified Tables), which is no longer extant in its entirety but is widely quoted by later astronomers. The initial observations of the Shammāsiyya observatory were evaluated and deemed unsatisfactory; it was only then that Al-Ma'mūn directed a team of astronomers to conduct additional observations at Mount Qāsiyūn in order to produce better results. At Mount Qāsiyūn, observations were made daily for at least one whole year and intermittently in the following two or three years.

Although not all of the institutional features were fully developed, the observatories of Al-Ma'mūn were unambiguously endowed with many of the characteristics of formal administrative and scientific institutions. These include the adoption of a set program of research, the choice of fixed locations appropriate for conducting the desired observations, recruitment of specialized staff, and the collaboration of several scientists in a research undertaking. In two separate instances, groups of professionals were assembled in special locations to conduct astronomical research. Directors were appointed for each of these groups, and attempts were made to organize the finances of the projects and pay the salaries of the staff. Scientific organization also involved identifying the things to be observed and conducting simultaneous observations of the same astronomical phenomenon in more than one location. In short, this was a coordinated, collective effort, undertaken by specialized professionals over a long period of time for the purpose of fulfilling a set program of research.

Records of observational activities abound from the time of the observations of Shammāsiyya and Qāsiyūn onward. Although many observations in the ninth and tenth centuries were private, they were often extremely significant from the point of view of advancing scientific activity.[62] In contrast to private observatories, which tended to use small instruments, official observatories that enjoyed financial support as well as a long, continued existence tended to use larger instruments, which had the advantage of producing more accurate observations.[63]

An important development took place in eleventh-century Iṣfahān, where a large and highly organized observatory was established under the patronage of the Saljūq ruler Malikshāh. The observations in this observatory were planned over a period of thirty years, which is the time taken by Saturn, the farthest planet from the earth, to complete one full revolution. Although the observatory functioned for only eighteen years and was shut down when its founder died, it was the first official observatory to last for such a long period of time, which gave it the status of a long-lived scientific institution.

The most famous institutionalized observatory was established in Marāghā in the thirteenth century under the patronage of the Ilkhānid Hulāgu and under the directorship of Naṣīr al-Dīn al-Ṭūsī (d. 1274). It was built on a large piece of land and was financed by assigning waqf revenues to support it. Because of its financial autonomy, the observatory was able to survive after the death of its founder and was active for more than fifty years. The Marāghā observatory served as a center of astronomical research and attracted a large team of astronomers from all over the Muslim world. These were the most talented astronomers of the time, and their collaboration, despite their diverse regional backgrounds, is a compelling illustration of the mobility of scientists and the universality of Arabic scientific culture. The Marāghā astronomers engaged in various kinds of scientific research, built specialized observational instruments, compiled new tables (the *Ilkhānid Zīj*), and did advanced work on planetary theory. The Marāghā institution served as a model for the organization of the large fifteenth-century Ulegh Beg observatory at Samarkand, the sixteenth-century Taqī al-Dīn observatory at Istanbul, and the eighteenth-century Jai Singh observatory at Jaipur.

A different model for the social sanctioning of astronomy and the related integration of astronomical institutions within society is shown by the emergence of the office of the timekeeper in many mosques all over the Muslim world. Although timekeeping activities were not identical to the activities conducted at observatories, they did provide stable resources for use by individual astronomers. In some cases, the achievements of timekeepers were scientifically superior to those of observatory astronomers. The best example is that of the celebrated fourteenth-century astronomer Ibn al-Shāṭir (d. 1375), who occupied the position of timekeeper at the Umayyad mosque in Damascus. Ibn al-Shāṭir conducted extensive observations, designed and constructed new instruments, and produced advanced contributions to Islamic astronomy in the field of planetary theory. Although there is no information on the observational setup utilized by Ibn al-Shāṭir, his observations were of great importance from a scientific point of view, and some of his theoretical contributions were based on his observations. It is clear, therefore, that Ibn al-Shāṭir's observational activity, most probably conducted within the facilities of the timekeeper's office at the Umayyad mosque, were at least as significant as those conducted at formal observatories. It is thus reasonable to think of the place where a timekeeper did his work as a minor observatory, one that, in contrast to formal observatories, was more stable, financially independent, and fully integrated into traditional social institutions.

The Consequences of Hybridity

The practice of science did not occur in a vacuum, as is already clear, and while the modes of supporting scientific practice changed over time and varied with discipline, there can be no doubt that, right from the beginning, the sciences were an enduring feature of the intellectual landscape of Muslim societies. The beginnings of scientific practice were instrumental in shaping later scientific practice in ways that had lasting effects on even later developments. In the remainder of this chapter, I will give a brief overview of some of the salient characteristics of this scientific practice.

The predominance of the Greek scientific tradition and its vital influence on the development of the Arabo-Islamic sciences is a given. But Arabic science was not a mere museum of Greek scientific knowledge. Arabic scientists did more than simply preserve the Greek scientific legacy and pass it on to its European heirs. In fact, evidence from the earliest extant scientific sources indicates that the translation movement was concurrent with, rather than a prerequisite for, scientific research in the Islamic world. Simultaneous research and translation took place in more than one field, and in more than one instance, when some of the scientific texts were being translated, they were also being reformulated and transformed.

It is worth keeping in mind as we attempt to explain this

phenomenon that the absorption of the Greek, Persian, and Indian scientific legacies into Islamic culture took place in the context of a dominant Islamic polity. To put it differently, Muslims were not in a position of political subordination, which explains the ease with which they borrowed from other cultures. More important than this sense of security, however, was the multiplicity of the scientific legacies that informed Islamic science. In a classic and much-quoted work on the history of Arabic astronomy published in 1911, Carlo Nallino downplays the Persian and Indian influences and suggests that the scientificity of Arabic science was largely due to its reliance on the Greek sciences.[64] Without denying the disproportionate weight of the Greek tradition, I would argue that the other two legacies were extremely important in two respects. First, they were important because they explain some of the earliest research in fields such as algebra that were concurrent with, and not subsequent to, the first translations of Greek mathematical works.[65] And second, they were important because the availability of multiple scientific traditions to choose from allowed an eclectic and discriminating approach to each of the scientific legacies. The early dynamics of translation and original composition suggest that the small number of non-Greek translations had the effect of hybridizing knowledge, which in turn had a lasing impact on the attitude toward the dominant Greek sciences.

Let me illustrate by again using the example of algebra. The first work in history to consider algebraic expressions irrespective of what they may represent was *Kitāb al-Jabr wal-Muqābala* by Muḥammad ibn Mūsā al-Khwārizmī (fl. 830). This book was written in the first quarter of the ninth century and was rightly considered by Arab mathematicians, as well as by early and late historians, as a pivotal achievement in the history of mathematics. Al-Khwārizmī himself was aware of the novelty of his work:

he used a title never used before in earlier disciplines, and he provided a new technical terminology without much parallel in earlier traditions. The perceived objectives of Al-Khwārizmī's work were equally original: to provide a theory for the solution of all types of linear and quadratic equations by using radicals and without restricting the solution to any particular problem. The subject of Al-Khwārizmī's new discipline was equations and roots: all geometric and arithmetic problems were reduced, through algebraic operations, to normal equations with standard solutions.[66]

The work of Al-Khwārizmī was only the first in a long and increasingly more sophisticated tradition of algebraic research. Almost immediately after the emergence of this new field, other mathematicians started developing it and exploring the possibilities for applying it to other mathematical disciplines. In the tenth and eleventh centuries, a new research project introduced by Al-Karajī (late tenth century) focused on the systematic application of the laws of arithmetic to algebraic expressions. Already in the second half of the ninth century, Qusṭā ibn Lūqā (d. 912) had translated the first seven books of the *Arithmetica* of Diophantus into Arabic. Significantly, however, the Arabic translation was given the title *The Art of Algebra*. By using the language of the new field of algebra, the translator reoriented the *Arithmetica* of Diophantus and provided instead an algebraic interpretation of the arithmetic. In this instance, the translation from Greek into Arabic was both motivated and conditioned by the earlier original research in Arabic algebra. Thus, the Greek arithmetic that Al-Karajī applied to algebra had already been modified, even as it was being translated, under the influence of the work of Al-Khwārizmī and his successors.[67] Al-Khwārizmī's work was considerably less sophisticated than later Arabic algebra in diverse traditions, but its influence was monumental. It also illustrates the transformative effects of the cross-application of multiple scientific disciplines and the way old sciences can be restructured and new sciences constructed by applying the tools of one discipline to another.

The formative effects of working from multiple traditions are also shown in astronomy. The first astronomical texts that were translated into Arabic, in the eighth century, were of Indian and Persian origin. The earliest extant Arabic astronomical texts date to the second half of the eighth century. Two astronomers, Muḥammad ibn Ibrāhīm al-Fazārī and Yaʿqūb ibn Ṭāriq, are known to have translated an eighth-century Indian astronomical work known as *Zīj al-Sindhind.* Sources indicate that they produced this translation after 770, under the supervision of an Indian astronomer visiting the court of the ʿAbbāsid caliph Al-Manṣūr (r. 754–75). Extant fragments of the works of these two astronomers reveal a somewhat eclectic mixing of Indian parameters with elements of Persian origin and some elements from the Hellenistic pre-Ptolemaic period. These fragments also reflect the use of Indian calculation methods and the use of the Indian sine function in trigonometry, instead of the cumbersome chords of arc used in Greek astronomy. Late Arabic sources also contain references to the *Zīj al-Shāh,* a collection of astronomical tables based on Indian parameters, which was compiled in Sasanid Persia over a period of two centuries.[68]

The first original work of Arabic astronomy still extant is Al-Khwārizmī's *Zīj al-Sindhind* (which is unrelated to the translation of the Indian text with the same name). It contains tables for the movements of the sun, moon, and five planets with introductory remarks on how to use the tables. Most of the parameters used by Al-Khwārizmī are of Indian origin, but some were derived from Ptolemy's *Handy Tables,* and no attempt was made to harmonize the two sources. This book is significant not only because of its content but also because of its timing: it was written at the same time that Ptolemy's *Almagest*, the most influential astronomical work of antiquity, was first translated into Arabic. The introduction of Ptolemaic astronomy into Arabic science thus occurred in the context of two significant trends. First, research in Arabic astronomy was proceeding hand in hand with translation; despite the manifest superiority of Ptolemaic astronomy, it did not exclusively set the agenda for future research in Arabic astronomy. The second trend was the selective use of parameters, sources, and methods of calculation from different scientific traditions. As a result, the Ptolemaic tradition was made receptive right from the beginning to the possibility of observational refinement and mathematical restructuring. These revisionist tendencies characterize the first period of Arabic astronomy.

Similar trends can be traced in medicine. The most famous of the early translators and physicians are Yūḥannā ibn Māsawayh (d. 857), head of the ʿAbbāsid library Bayt al-Ḥikma, and Ḥunayn ibn Isḥāq (d. 873), who, along with his students, translated almost all of the then known Greek medical works into either Syriac or Arabic. While these translations were being made, original works were being composed in Arabic. Ḥunayn, for example, composed a few medical treatises of his own; of these, *Al-Masāʾil fī al-Ṭibb lil-Mutaʿallimīn* (Questions on Medicine for Students) and *Kitāb al-ʿAshr Maqālāt fī al-ʿAyn* (Ten Treatises on the Eye) were influential and quite original. They include very few new observations; rather, their creativity lies in their new organization and, in the case of the second book, in the deliberate attempt to exhaust all questions related to the eye. A solid command of medical knowledge was needed to produce these works.

The most famous work of the early period of Arabic science

was composed by 'Alī ibn Sahl Rabbān al-Ṭabarī (fl. mid-ninth century), a Christian convert to Islam from Marw. Al-Ṭabarī's book *Firdaws al-Ḥikma* (Paradise of Wisdom) was the first comprehensive work of Arabic medicine to integrate and compare the various medical traditions of the time. A section on Indian medicine provides valuable information on its sources and practices. The author adopted a critical approach to enable a choice between different practices.

Indian medicine was far less crucial than Hellenistic medicine in shaping the Arabic medical tradition, although occasionally physicians would compare Greek and Indian medicine and opt for the latter; this was the case with Al-Samarqandī in the eleventh century, but he was an exception. The main role of Indian medicine was not to define the contours of the Arabic medical tradition but to set the tone for some of its initial interests.[69] Although the Greek scientific legacy was the dominant one, a mere awareness of more than one available tradition encouraged a critical and selective approach that pervaded all fields of Arabic science.

The Critical Outlook

A critical awareness of the possibility of multiple interpretations and approaches to various scientific disciplines characterized the practice of science during the ninth century and beyond and resulted in three related trends: a thorough examination of the details of various disciplines; an attempt to systematize and generalize these disciplines; and the production of exhaustive syntheses of individual disciplines and related ones. The following examples from astronomy, medicine, and optics illustrate these trends.

Astronomy. The *Almagest* was and is rightly considered the main authoritative work of antiquity that deals with astronomy. In this book, Ptolemy synthesizes the earlier knowledge of Hellenistic astronomy in light of his own new observations. The main purpose of the book was to establish the geometric models that would accurately account for celestial observations. A large part of the book is dedicated to the methods for constructing various models and for calculating the parameters of these models. Ptolemy also provides tables of planetary motions to be used in conjunction with his models. Of all the books of antiquity, the *Almagest* represents the most successful work of mathematical astronomy: its geometric representations of the universe provided the most accurate and best predictive accounts of celestial phenomena.[70]

A significant part of the intensive astronomical research of the ninth century was dedicated to the dissemination of Ptolemy's astronomy, not just by translating parts or all of his work but also by composing summaries and commentaries on it. Ptolemy was thus made available and accessible to a large audience among the educated classes. In the first half of the ninth century, Al-Farghānī, for example, wrote his *Kitāb fī Jawāmiʿ ʿIlm al-Nujūm* (A Compendium of the Science of the Stars), a book that was widely circulated in both the Arabic version and later Latin translations. It provides a brief and simplified descriptive overview of Ptolemaic cosmography without mathematical computations. Unlike the *Almagest*, however, it starts with a discussion of calendar computations and conversions between different eras. Although Al-Farghānī's primary purpose was to introduce Ptolemaic astronomy in a simplified way, he also corrected Ptolemy. Using the findings of earlier Arab astronomers, he gives revised values for the obliquity of the ecliptic, the precessional movement of the apogees of the sun and the moon, and the circumference of the earth. At the start of the ninth century, earlier astronomers had already taken a critical approach, although it was restricted to the correction of constants and parameters.[71]

Right from the beginning, then, Arabic astronomers set out to rectify and complement Ptolemaic astronomy. Having noted several discrepancies between new observations and Ptolemaic calculations, Arab astronomers reexamined the theoretical basis of Ptolemy's results. This critical reexamination took several forms. One example of the critical works of the ninth century is *The Book on the Solar Year*, which was wrongly attributed to Thābit ibn Qurra (d. c. 901) but was produced around his time. Ptolemy's precession constant is corrected in the book, and although Ptolemy's geometrical representations are retained, the author questions his observations and calculations. The book also presents proof that the apogee of the solar orb—the celestial sphere that carries the sun—moves relative to the ecliptic in connection with the precessional movement of the sphere of the fixed stars, just as the apogees of the other planets and the moon do. Other astronomers devised enhanced methods of calculation. New mathematical tools were also introduced to modernize the computational procedures. In *Al-Zīj al-Dimashqī*, written around the middle of the ninth century, Ḥabash al-Ḥāsib introduces sines, cosines, and tangents, trigonometric functions unknown to the Greeks. Ḥabash also worked on a problem not treated in the Greek sources; he examined the visibility of the crescent moon and produced the first detailed discussion of this complicated astronomical problem. He not only undertook to verify the results of the *Almagest* but also expanded these results and applied them to new problems. Although the general astronomical research of this period was conducted within the framework of Ptolemaic astronomy, he and others reworked and critically examined the Ptolemaic observations and computational methods and, in a limited way, were even able to explore problems outside that framework.

Thābit ibn Qurra was one of the main scientists of the ninth century. A pagan from Ḥarrān whose birth language was Syriac and whose working language was Arabic, although he was also fluent in Greek, Thābit joined the Banū Mūsā circle of scientists in Baghdad and produced numerous works on several scientific disciplines. Of about forty treatises on astronomy, only eight are extant. All reflect Thābit's full command of Ptolemaic astronomy and illustrate the level to which this astronomy was thoroughly absorbed by Arab astronomers. A couple of the extant astronomical treatises are of particular interest. In one, Thābit analyzes the motion of a heavenly body on an eccentric orbit, and the model he uses is Ptolemaic. Yet where Ptolemy describes without giving proof, Thābit provides a rigorous and systematic mathematical proof with the aid of the theorems of Euclid's *Elements*. In the proof, Thābit introduces the first known mathematical analysis of motion. For the first time in history, as far as we know, he also refers to the speed of a moving body at a given point. In another work, Thābit provides general and exhaustive proofs for problems that Ptolemy examines only for special cases or for boundary conditions. Thābit also devotes a work to lunar visibility. His solution, which is far more complex than Ḥabash's, exhibits the same mathematical rigor everywhere apparent in his work: he proves the general law that applies to the visibility of any heavenly body, then applies this law to the special case of the crescent moon. The work of Thābit is significant because it illustrates the high creativity of Arabic astronomy in its earliest period, the roots of which lie in the application of diverse mathematical disciplines to one another. Cross-application had the immediate effect of expanding the frontiers of disciplines and introducing new scientific concepts and ideas, and the use of systematic mathematization transformed the methods of reasoning and enabled further creative developments in the branches of science.[72]

Already in the ninth century, then, Arabic astronomy integrated all available knowledge from earlier traditions and was positioned to add to it. The achievements of the ninth century laid the foundation for the high-quality work in the following two centuries. In the tenth and eleventh centuries there were important developments in the science of trigonometry, which had dramatic effects on the accuracy and facility of astronomical calculations. The earlier examinations of Ptolemaic astronomy led to systematic projects that, rather than addressing the field in its totality, focused on specific aspects of astronomy. One of the main characteristics of this period was the tendency to provide exhaustive synthesizing works on particular astronomical topics. The work of ʿAbd al-Raḥmān al-Ṣūfī (who was born in Rayy, worked in Shīrāz and Iṣfahān, and died in 986) illustrates this tendency. In his famous book *Kitāb Ṣuwar al-Kawākib al-Thābita*, Al-Ṣūfī reworked the star catalogue of the *Almagest* on the basis of a corrected value of one degree every sixty-six years for the precessional movement of the stars and incorporates several other new observations and verifications, producing an accurate representation of the constellations and their coordinates and magnitudes. Another example of the tendency to synthesize is *Al-Zīj al-Ḥākimī al-Kabīr*, a monumental work in eighty-one chapters of which only about a half are preserved. The book, by Ibn Yūnus (Cairo, d. 1009), is a complete treatise on astronomy containing tables on the movement of the heavenly bodies, their various parameters, and instructions on the use of the tables. Here, too, the objective of the work is to provide an exhaustive documentation of previous observations, subsequent verifications or corrections of these, and new observations recorded by the author. The synthesizing trend culminated in Abū al-Rayḥān Muḥammad ibn Aḥmad al-Bīrūnī's (973–c. 1048) *Al-Qānūn al-Mas'ūdī*, a synthesis of the Greek, Indian, and Arabic astronomical traditions.[73]

Advances in trigonometry resulting from the full integration of Indian achievements in the field, as well as from new discoveries in the tenth and eleventh centuries, played a central role in the development of Arabic astronomy. This tendency was itself part of the systematic mathematization of disciplines, which contributed to the expansion of their frontiers. Equipped with new and more rigorous mathematical tools, Al-Bīrūnī, like many of his predecessors and contemporaries, provided exhaustive studies of specialized topics within astronomy. His treatises cover such topics as shadows; the theory, construction, and use of astrolabes; and the coordinates of geographical locations. In most of these monographs, Al-Bīrūnī starts with a thorough critical overview of older theories and mathematical methods for solving the particular problems in question, then proceeds either to choose one or to propose his own alternative theory. Al-Bīrūnī's work represents a critical assessment of the state of the science of mathematical astronomy up to his own time; like other such comprehensive surveys, it in effect exhausted the possibilities of expanding the astronomical disciplines from within. To achieve progress, scientists needed to move in new directions, devise new strategies, and explore new programs of research.

Medicine. In the tenth and eleventh centuries, as with astronomy, Arabic medicine exhibited tendencies toward empirical precision, exhaustiveness, and systematization. *Al-Ḥāwī fī al-Ṭibb* (The Comprehensive Book on Medicine) by Abū Bakr al-Rāzī (d. 925) is an example. It is a huge, comprehensive work that, in a current incomplete copy, falls into twenty-three volumes. The book is not organized according to formal theoretical paradigms; rather, it is a compendium and encyclopedia of clinical medicine, including all earlier writings on diseases and treatments known to Al-Rāzī, as well as his own clinical observations. In several places, Al-Rāzī criticizes Galen, the preeminent second-century Greek physician, pointing out that his own clinical observations do not conform with Galen's assertions. Another attempt to provide a synthesis of all available medical knowledge was the thirty-volume medical encyclopedia *Kitāb al-Taṣrīf li man ʿAjiza ʿan al-Taʾlīf* by the Cordovan scholar Abū al-Qāsim al-Zahrāwī (d. 1013). The bulk of this work deals with symptoms and treatment, reflecting once again the increased interest among many medical scholars in clinical comprehensiveness.

Paralleling these attempts at synthesis were attempts at rigorous systematization. For the first time, scholars tried to organize the vast body of medical knowledge in all of the branches of medicine into one logical structure. In the tenth century, ʿAlī ibn ʿAbbās al-Majūsī wrote his *Kitāb al-Kāmil fī al-Ṣināʿa al-Ṭibbiyya* (The Complete Book of the Medical Art), also known as *Kitāb al-Malakī* (The Royal Book), with the explicit intention of applying a theoretical framework that would provide medicine with a structure and an organizational principle. Although Al-Majūsī's book served as a popular handbook of medicine, it was soon replaced by what became the single most influential book on theoretical medicine in the Middle Ages and up to the seventeenth century. This was *Al-Qānūn fī al-Ṭibb* (The Canon of Medicine) by the celebrated Muslim philosopher and physician Ibn Sīnā (d. 1037). Ibn Sīnā's magnum opus was written, as the title indicates, with the intention of producing the definitive canonical work on medicine in terms of both comprehensiveness and theoretical rigor. In the book, Ibn Sīnā provides a coherent and systematic theoretical reflection on inherited medical knowledge, starting with anatomy and continuing with physiology, pathology, and therapy. Although he includes many bedside observations and a few original contributions of a purely practical nature, Ibn Sīnā's main achievement is not primarily in the clinical domain. Rather, he produced a unified synthesis of medical knowledge that derived its coherence from the relentlessly systematic application of logic and theoretical principles.[74]

Optics. The Arabs acquired a large body of Hellenistic optical knowledge that covered the physical as well as the geometric study of vision, the reflection of rays on mirrors (catoptrics), burning mirrors, and such atmospheric phenomena as rainbows. Within two centuries, the field of optics was radically transformed, and Arabic optics acquired the characteristics of a new field of study with distinct methodologies and approaches. The first Arabic compositions on optics were in the eighth century. In the ninth came the works of Yūḥannā ibn Māsawayh, Ḥunayn ibn Isḥāq, Qusṭā ibn Lūqā, and Thābit ibn Qurra. Although ninth-century works primarily addressed physiological optics, they also treated other subjects, in separate studies: burning mirrors, mirror reflections, and geometrical and physical optics. Abū Isḥāq al-Kindī (d. c. 873) is said to have produced ten treatises on geometrical and physical optics, of which at least four are extant. Both Qusṭā ibn Lūqā and Al-Kindī adopted a deliberate strategy in their research. To rectify the results in one particular subfield of Hellenistic optics, they drew on other subfields, with the intention of combining the geometry and the physiology of vision.

The most important work of Arabic optics is undoubtedly *Kitāb al-Manāẓir* by Ibn al-Haytham (d. 1039). The creative expansion of optical research in all of its subfields reached a peak under Ibn al-Haytham, who covered in an integrated research project all of the traditional themes of optics as well as those invented by his Arabic forerunners. This project effectively undermined not only the premise but also the structure of Greek optical research. At a basic conceptual level, Ibn al-Haytham rejected the Hellenistic theories of vision and introduced a radically different theory. Vision, according to earlier theories, was a result of contact between the eye and the object either through a ray emitted from the eye to the object—as in the extramission theories of Euclid and Ptolemy—or through the transmission of a "form" from the object to the eye, as in the intromission theories of Aristotle and the atomists. Ibn al-Haytham's remarkable insight was that what is sensed is not the object itself; rather, an image of the object is formed as a result of the reflection of light from the object to the eye. Ibn al-Haytham could thus proceed to study the geometric aspects of the visual cone theories without having to explain at the same time the psychology of perception. He also benefited from advances in the study of the physiology of the eye, thus, in the end, integrating into his theory of vision the cumulative results of mathematical, physical, and medical research.[75]

The Invention of New Sciences

A distinctive result of the reorganization and cross-application of different sciences to one another was the invention of new sciences. Algebra, as we have seen, was conceived as a new science with a distinct subject matter, technical terminology, methods, and even name. The significance of Al-Khwārizmī's

Kitāb al-Jabr wal-Muqābala, in addition to its originality, was in the research it triggered when applied to other branches of mathematics. Starting with Al-Khwārizmī and continuing with Al-Khayyām (1048–1131) and Al-Ṭūsī, several mathematicians were fully aware of the utility of the cross-fertilization of mathematical disciplines and the novelty of this sort of research. They concocted unfamiliar titles for their books, coined technical terminology unique to their disciplines, organized their works in decidedly novel ways, and invented mathematical algorithms to solve the problems of their disciplines; above all, they came up with totally new subjects and mathematical concepts. They did not merely restructure Hellenistic mathematical knowledge but created new mathematical disciplines. Such innovations were made possible by the deliberate and systematic application of three mathematical disciplines to each other: algebra, arithmetic, and geometry.

I have already mentioned Al-Karajī's systematic application of the laws of arithmetic to algebraic expressions. After Al-Karajī, the central efforts in algebraic research focused on the arithmetization of algebra, a genre of research that was new in both content and organization. While the application of arithmetic to algebra occupied center stage in algebraic research, the theory of algebraic equations also continued to develop. Thābit ibn Qurra provided systematic geometrical interpretations of algebraic procedures and explained quadratic equations geometrically. Other mathematicians attempted to do the reverse and explain geometrical problems in algebraic terms. Aware of the difficulty of solving cubic equations by using radicals and demonstrating such solutions geometrically, Al-Māhānī (ninth century) introduced the first algebraic formulation of a solid problem. Mathematicians then increasingly resorted to conic sections to solve cubic equations that could not be solved with radicals. Unlike earlier attempts to demonstrate geometrically equations whose roots are known through algebraic solutions, the objective of this last research was to find, with the help of geometry, the roots of equations that are not solvable numerically. A continuous tradition of partial contributions to this field began in the ninth century and culminated in the systematic work of Al-Khayyām, who elaborated a geometrical theory for first-, second-, and third-degree equations. For all types of third-degree equations, he provides a formal classification according to the number of terms, then solves the equations by means of the intersection of two conic sections.[76]

Trigonometry was another hybrid mathematical discipline in which Arab scientists enriched and eventually reoriented earlier scientific knowledge. It was initially developed in conjunction with research in astronomy, but it became a mathematical discipline in its own right. Ptolemy's astronomy, it will be recalled, was superior in its models, but it rested on elementary geometrical propositions. Ptolemaic astronomical computations were based on a single function, the chord of a circular arc, and the only tool for spherical computation was the Menelaus theorem, a cumbersome formula indicating the relationship between the six segments that result from the intersection of four arcs in a complete quadrilateral. Soon after translating Ptolemy and adopting his models, Arab astronomers augmented his geometry with the powerful sine function of Indian trigonometry and, in the ninth century, introduced the tangent function. The emergence of trigonometry as an independent science, however, required two additional developments: first, identifying the spherical triangle as the object of study as opposed to the calculus of chords on the spherical quadrilateral; and second, including the angles of triangles in the refocused calculus and not just the sides. The first accounts of the spherical triangle appeared by the end of the tenth century. Already in the eleventh century, all six relations of the right-angled triangle appeared in various texts, including, among others, *Maqālīd 'Ilm al-Hay'a* by Al-Bīrūnī. In the thirteenth century, Al-Ṭūsī wrote the first treatise on trigonometry without reference to astronomy, thus confirming the creation of another independent discipline.

The science of weights developed in much the same way. The Arabs inherited a number of Hellenistic theoretical studies on geometrical statics, including, among other subjects, the mathematical study of the laws of equilibrium, the concept of a center of gravity, and hydrostatic studies of the equilibrium of bodies in liquids. They also inherited practical studies of simple machines for lifting and moving bodies. As in the other sciences, one of the main trends in Arabic statics was the systematic use of inherited as well as new mathematical techniques, especially algebra. This enabled both a generalization of Greek statics and the invention of new fields within the discipline. In particular, the use of a dynamic approach to the study of statics—itself a result of the systematic application of new mathematical skills—led to the emergence of the science of weights, which provided the new theoretical foundation for the science of mechanics.[77] The application of this dynamic approach to hydrostatics, as reflected in Al-Khāzinī's study of the motions of bodies in fluids, led to the emergence of the new field of hydrodynamics.[78]

Equally important developments came in the field of pharmacology. In the twelfth century, several encyclopedic works on pharmacology were compiled by such scientists as Abū Ja'far al-Ghāfiqī and Abū al-'Abbās al-Nabātī. The culminating work was the great synthesis by Ibn al-Bayṭār (d. 1248), *Al-Jāmi' li-Mufradāt al-Adwiya wal-Aghdhiya* (The Dictionary of Simple Medicines and Foods). The most complete treatise of applied botany produced in the Middle Ages, it draws information from more than 150 sources and lists more than 2,000 simples in alphabetical order. Ibn al-Bayṭār brings together all the accumulated knowledge of the numerous inherited traditions and adds to it his own knowledge and experiences. He gives the names of simple medicines in all the written languages known to him, including several local dialects. He also succeeds in finding the Arabic names of almost all of the simples listed in the work of Dioscorides (first century CE). Ibn al-Bayṭār's method of research is as significant as the results of the research. Following the lead of Dioscorides and several Arab botanists, among them his own teacher, Ibn al-Bayṭār traveled to North Africa, Greece, Anatolia, Iran, Iraq, Syria, Arabia, and finally Egypt, conducting onsite research in all these places. He settled in Egypt, where he was appointed head pharmacist of the country, but he continued to conduct field trips to Syria to collect new data and verify earlier findings. Like many other contributions in this field, Ibn al-Bayṭār's *Al-Jāmiʿ* illustrates at once the tendency to synthesize and the tendency to rely on observation for the expansion of scientific knowledge.[79]

Ibn al-Bayṭār's descriptions are extremely accurate, but the primary purpose of his book was medical. Another work, by Abū al-ʿAbbās al-Ishbīlī (d. 1239) entitled *Al-Riḥla al-Mashriqiyya* (The Eastern Journey), has a purely botanical focus. The book is lost, but Ibn al-Bayṭār quotes it in full in more than one hundred entries. Unlike works that contain botanical information but whose authors were ultimately interested in the medical use of plants, Al-Ishbīlī's book expressed his purely botanical interests. Although he was a famous physician, Al-Ishbīlī provides meticulous descriptions of plants as plants, not as medicines. His work illustrates the familiar process, seen in the subfields of mathematics, through which new disciplines emerged as a result of the expansion and systematization of older ones.

The Epistemological Rehabilitation of Practical Knowledge

The invention of numerous new scientific disciplines did not result solely from the cross-fertilization of different theoretical sciences but owed much as well to the attitude toward practical skills and knowledge that derived from them. The Greek attitude toward crafts and practical knowledge is best expressed in a reference in Plutarch's *Lives* to Archimedes, whose machines helped destroy Marcellus's fleet when it was attempting to capture Syracuse. Plutarch says: "These machines he had designed and contrived, not as matters of any importance, but as mere amusements in geometry; in compliance with King Hiero's desire and request, some little time before, that he should reduce to practice some part of his admirable speculation in science, and by accommodating the theoretic truth to sensation and ordinary use, bring it more within the appreciation of the people in general." In the same reference, Plutarch talks about "Plato's indignation" over practical crafts and his invectives against them as "the mere corruption and annihilation of the one good of geometry, which was thus shamefully turning its back upon the unembodied objects of pure intelligence to recur to sensation, and to ask help (not to be obtained without base supervisions and depravation) from matter; so it was that mechanics came to be separated from geometry, and, repudiated and neglected by philosophers, took its place as a military art."[80]

With the exception of some staunch Hellenists, people in Muslim societies took an attitude toward practical crafts and knowledge that was fundamentally different from the Greeks'. In a culture that ascribes the highest value to the legal religious sciences, which are quintessentially practical, there was little room for a negative attitude toward practical sciences of any sort. Right from the beginning, utilitarian considerations provided incentives for the pursuit of science. Practical needs included land surveying, inheritance algebra, irrigation technologies, calendar computations, and timekeeping, in addition to medicine and related sciences. Some of the earliest scientists and patrons of science—such as the Banū Mūsā brothers (ninth century), leading members of the social and political elite affiliated with the ʿAbbāsids—were involved in public projects in which practical scientific knowledge could be deployed, and also commissioned a large number of translations and composed works in such fields as mathematics. In a telling indication that no distinction was made between practical and theoretical knowledge, the earliest extant book on automatic machines was composed by the Banū Mūsā.[81]

The list of practical disciplines officially recognized as sciences is long, and it provides further evidence for a distinct Islamic attitude toward practical knowledge. These disciplines had separate, often completely new titles; scientists did not shy away from composing numerous books on these subjects; they often defended the epistemological value of their knowledge; and the classifications of science genre that proliferated in Muslim societies almost invariably listed these practical disciplines as full-fledged sciences. Below are a few examples.

The science of mechanical devices (*ʿilm al-ḥiyal*) drew on a rich Greek tradition, including the work of Archimedes. What is noteworthy in the Arabic tradition is the unapologetic promotion of the use of technical lore to the level of science and then its development into what, for all practical purposes, was mechanical engineering. Early descriptions of mechanical devises were strictly schematic: diagrams were used to illustrate the theory underlying a device, not to provide information on its construction or dimensions. The first book that can be considered a mechanical engineering handbook was *Kitāb fī Ma'rifat al-Ḥiyal al-Handasiyya* by Al-Jazarī (c. 1200).[82] In addition to providing schematic illustrations of how machines work, the book gives detailed instructions on the dimensions of their various parts, the materials to be used and their treatment (for example, lamination to prevent wrapping), casting techniques, and information on finishing, calibration, and priming procedures. In short, the book provides all the information needed to manufacture particular machines—not only to understand the way they work.

Some elaborate and detailed guidelines for the application of technological know-how are preserved in the field of irrigation engineering. Numerous irrigation methods were inherited from the ancient agrarian-based Near Eastern societies. The Muslims in turn supported intensive agricultural development projects. Arabic chronicles provide considerable information on the scale and significance of projects initiated by the Umayyads, the 'Abbāsids, and others. These projects were crucial for developing the economies of Muslim societies and providing for the needs of newly established cities or expanded older ones. Many irrigation projects were massive in scale: they often involved the building of dams to control and regulate the flow of rivers, the use of water-raising machines to transfer water for irrigation and water supply, and the building of extensive networks of canals and *qanāt*s to divert water to places where it was needed. Highly advanced technical and administrative skills were needed to construct and manage large-scale irrigation and water supply projects. One Arabic chronicler reports that a tenth-century supervisor of the irrigation system of the city of Marv and its environs was in charge of the more than ten thousand workers who built, maintained, and controlled the system. Specialized technical skills—for surveying and excavating canals, for example—were also developed in connection with irrigation. Several treatises, written on quantity-surveying methods, provided detailed instructions for the management of the construction of large-scale irrigation systems.

The agricultural sciences were pursued with great success in Al-Andalus (now Spain). In the numerous works written on agronomy, scientists emphasize relying on practical knowledge in deriving the principles (*mabādi'*) of this science. Those who rely solely on the pure philosophical tradition are pejoratively referred to as imitators (*muqallidūn*) because they do not ground their assertions on knowledge derived from practice (*tajrīb*). In many instances where the flaws of the philosophical agricultural tradition are pointed out, authors insist that to have sound foundations, the science of agriculture ought to combine theoretical knowledge with knowledge that derives from manual work, on the one hand, and with experimentation that provides definitive proof (*burhān*), on the other hand. In one source, the ideal specialists in the science of agriculture are called philosopher-peasants (*hukamā' al-fallāḥīn*).[83]

In the mathematical sciences and astronomy, several practical disciplines were established and became permanent features of the scientific landscape. Practical astronomical problems occupied numerous astronomers who were responsible for significant advances in both theoretical and practical branches of the field. Some of these problems had to do with the practical needs of society, such as finding the location of one place with respect to another, which requires determining longitudes and latitudes as well as invoking other aspects of mathematical geography. Other, specifically Islamic problems were related to worship: determining the times of prayer, the times of sunrise and sunset (related to times of fasting), the direction of the qibla, and the visibility of the crescent moon (to verify the beginnings of the lunar months). Calendar computations were also needed to determine the correspondence between the astronomical year and the rounded calendrical year. Everything from simple approximative techniques to complex mathematical ones was used to solve these problems. Complex theoretical analysis often went far beyond the initial scope of the examined problems. Many of the Islamic astronomical problems were treated in the new science of timekeeping (ʿilm al-mīqāt), whose status as a full-fledged science was never questioned.[84]

As we might have expected, medicine was another area where the practical and the theoretical intersected. An important focus in Arabic medicine was on expanding empirical medical knowledge—especially clinical or case medicine—and practical procedures for treatment, as opposed to making theoretical reflections on illness and health. One of the greatest representatives of this trend is Abū Bakr al-Rāzī. In his prolific writings, Al-Rāzī generated theoretical criticism of inherited medical knowledge. More important than his criticism, however, was his focus on method and practice. Throughout his works, Al-Rāzī stresses observational diagnosis and therapy more than he does the theoretical diagnosis of illnesses and their cures. He typically surveys all of the available medical knowledge, then provides a critical review on the basis of his own practice. His experience as a clinician was undoubtedly wide and rich, acquired in the course of a long career as the head of hospitals in Rayy and Baghdad. Some of Al-Rāzī's most original writings also derive from this lifelong work. His *Kitāb fī al-Jadarī wal-Ḥaṣba* is the first thorough account of the diagnosis of and treatment methods for smallpox and measles and the differences between their symptoms. What characterizes the book is its focus on clinical and not theoretical issues.

These and many other examples illustrate the tendency to correlate theoretical and practical knowledge and to treat crafts as sciences, thereby establishing a stable unity of theory and practice.[85] One effect of this promotion of practical knowledge was the increasing professionalization of science as a whole, which accompanied the increasing emphasis on the alliance of science with practical disciplines. In my view, one of the most important outcomes of this shift was the dramatic expansion of the consumer base of scientific knowledge. The tacit knowledge needed to understand the sciences was now shared by significantly larger sectors of the intellectual elite of Muslim societies. The sciences were now dealing with problems that punctuated every aspect of the daily lives of Muslims.

We have already seen in the example of the qibla how discussion was conducted at various levels, thereby allowing large groups of people to take part. The seemingly random use of old as well as new mathematical methods in the solution of astronomical problems also illustrates this trend. The same author may have used an archaic method in one place and an advanced method in another. Al-Bīrūnī, for example, used both the old, cumbersome Menelaus theorem and the new, elegant sine rule in several solutions of the problem of the direction of the qibla. Simultaneous use of different mathematical procedures can be attributed neither to the slow dissemination of scientific knowledge nor to the limited circulation of this knowledge. There is ample evidence of a high degree of mobility and efficient and speedy communication among scientists working in various regions of the Muslim world. Al-Bīrūnī himself did not travel to Baghdad, but he apparently corresponded with colleagues there and was fully aware of scientific developments there and elsewhere. Contrary to what the use of different methods may at first suggest, it was likely a result of the increasing diffusion of scientific knowledge among the educated elites. Within the broad ranks of these elites, "full-time" scientists were expected to keep up with the latest research in their fields, while scholars with a partial interest in science would be familiar only with older theories and methods. The use of a variety of mathematical methods thus indicates the degree to which scientific culture had filtered into society and the extent to which it had become available to average members of the educated class.[86]

Communities of Scientific Knowledge

Demonstrably, the culture of science struck deep roots in classical Muslim societies. Various developments contributed to the transformation of science from a peripheral, elitist activity (as it had been in earlier societies) to an institutionalized activity with an unprecedented scale of social support and participation. One aspect of the centrality of scientific culture was the existence of actual communities of scientists, who, as we have seen, had a sense of collective, professional identity. But more important than this collective sense of identity were the shared codes of practice, canons of study, and research agendas and projects, both empirical and theoretical, within the fields of specialization.

I underscore this point because so much earlier work on the history of Islamic science accepts that most of the original discoveries and contributions were isolated occurrences or happy guesses that had no impact on their Islamic environment and were appreciated only in Latin Europe. In the past few decades, this thesis has been largely undermined by the foundational research of published historians of Islamic science, but it continues to inform the discussions of general historians of science.[87] Now, research provides compelling evidence for the continuity and coherence of the Arabic scientific traditions.[88] I have already referred to the collaborative observational activities that started in the ninth century and continued unabated for many centuries thereafter.[89] Other examples can be found in the tradition of reforming Ptolemaic astronomy that started in the eleventh century and continued until at least the sixteenth and spanned most of the Islamic world. A definite and continuous research agenda guided the research of all the major theoretical astronomers, who read and commented on each other's work and in some cases even assembled to conduct joint research.[90]

Similarly, research on the disciplines of Arabic mathematics has revealed that for each instance of a seemingly isolated scientific breakthrough, there were in fact precedents and successors and an associated community of interested scholars and intellectuals. More than any other scholar, Roshdi Rashed has systematically explored the various mathematical sciences and established the continuities in each and every one of the disciplines. He illustrates, for instance, how Al-Khayyām's monumental contribution to the theory of algebraic equations was not isolated, as is often asserted in general surveys of the history of mathematics. Rashed shows how Al-Khayyām's work built on the earlier tradition of algebraic research and constituted only the beginning of a long and continuous tradition that was further transformed, half a century later, by Sharaf al-Dīn al-Ṭūsī (twelfth century). In its analytic approach, the work of Al-Ṭūsī on equations marked yet another beginning in the discipline of algebraic geometry: the study of curves by means of equations.[91]

Communities of scholars that include not only first-class mathematicians but also commentators of lesser reputation, as well as scholars working in other fields, contributed to the creation and diffusion of a multitude of mathematical traditions. For every celebrated scientist known to have conducted rigorous research in any field of science, many more practitioners provided the context without which such advances would have been impossible. We even have records of discoveries and inventions of mathematical theorems by more than one scholar at the same time. Al-Bīrūnī tells us, for example, that the general theorem of sines, known as the rule of four quantities, was discovered simultaneously and independently by astronomers from Khwārizm, Baghdad, and Rayy (Abū Naṣr ibn ʻIrāq, Abū al-Wafā' al-Būzjānī, and Al-Khujandī). Aside from claims of precedence, which reflect an active competitiveness among scientists, this account also suggests that the research needs of a particular science and the accumulated knowledge available to meet those needs make a discovery inevitable. In all the sciences, there always existed communities of scientists that provided temporal and spatial continuity for the culture of science, communities that practiced what Thomas Kuhn calls normal science. Normalized practices provided the social as well as the intellectual contexts for the exceptional moments of creativity in the history of Arabic science.

Although I have focused on some of the contexts in which an Arabo-Islamic scientific culture emerged and developed, and identified some common patterns that characterized the practice of science and the modes of scientific thinking, I recognize that historical continuity is not necessarily the same as epistemological continuity, and that there is no necessary connection between the historical formation of a science and its theoretical and epistemological structure. So, as important as contexts are, we need to accept that a context is patched together, and it has to be patched together—there is no way to fathom or imagine it fully. And since contexts are multiple at any given time and may conflict, I will not be confined to context in discussing Islamic science, although I will not dispense with it altogether. Rather,

I will focus on the complexities of texts, their internal logic, and their relations to various intellectual traditions. In doing so, I hope to describe, and to some extent explain, changes in the worldviews connected to and resulting from the historical practice of science in Muslim societies.

Chapter 2

Science and Philosophy

The relationship between science and philosophy in Islamic intellectual history is often described in one of two contradictory ways. Earlier historians of Islam and Islamic science repeatedly underscored the practical orientation of Islamic science and the lack of theoretical and philosophical rigor. Muslims, according to this thesis, viewed the sciences as crafts, not systems of knowledge, stripping them of their philosophical underpinnings and undermining their systematic and truly scientific nature. According to this view, the philosophical and scientific package that was imported and did not emerge from within Arabic society did not have a serious impact on Islamic culture; as a result, the whole question of the relationship between philosophy and science did not arise in Islam.[1]

The view of the Islamic sciences as mere crafts lacking theoretical coherence is perhaps the reason behind the notable discrepancy in most general histories of science between the detailed lists of the many technical contributions and the almost total absence of Islamic sciences from grand, integrative discussions of epistemological developments in the history of science.

When historians offer a conceptual analysis of epochal changes in the history of science, the cumulative legacy of the Islamic sciences is overlooked. It is seen as a rather mechanical continuation of the Greek one: the Islamic sciences are thought to have expanded and refined the Greek sciences without departing from them conceptually.[2] A justification for this oversight is seldom provided, but when one is, the reason given usually has to do with the role of philosophy or theory in science. To put it crudely, the authors assume that Islamic science was practical and hence theoretically or philosophically shallow.[3] The decline of Islamic science, according to this view, was a consequence of a lack of theoretical rigor.

In the past few decades, an alternative view has been proposed, often by historians of science as opposed to generalists. Instead of arguing that the Islamic sciences declined because of their feeble philosophical foundations, historians of astronomy in particular now argue that the motivation for the most important tradition of astronomical reform in the Muslim world was in fact philosophical.[4] Research in Islamic theoretical astronomy, the argument goes, was concerned primarily with cosmological or philosophical questions; "scientific" questions, in contrast, were presumably ones primarily involving observation and mathematics. Admittedly, this second view has not been used to explain large cultural trends in Islamic history, such as the decline of the sciences, but one result has been that some of the most competent historians of Islamic astronomy did not pay much attention to the philosophical and epistemological dimensions of the tradition they were studying.

In a sense, the two views regarding the role of philosophy in Islamic science echo, albeit in a rudimentary way, the debates over the function of scientific theory: whether it aims at describing and explaining reality or simply at describing and predicting perceived appearances of natural phenomena without offering natural or causal explanations. In the former case, science goes beyond appearances to explore causal connections, or, in the language of philosophers, "first causes."[5]

The question of science and philosophy was in fact raised in Islamic culture, the relationship between the two was reshaped in classical times, and this reshaping had an impact on the development of science. To explore this relationship, we need to distinguish between the generic use of the term "philosophy," which can be reasonably attributed to many sides of the debate, and philosophy as an inherited tradition and system of thought. It is in this latter sense that I will be referring to philosophy: as not just general philosophical thinking but a particular kind of philosophical thinking that has canons and traditions. Closely connected to this second sense of the word, though not identical with it, is philosophy as a professional practice or vocation. Thus, when tracing the relationship between science and philosophy we can either look at the relationship between systems of thought as expressed or implied in the writings of scientists and philosophers or at the relationship between scientists and philosophers as members of distinct professional groups. Again, without underestimating the significance of professional distinctions, I assert a primary interest here in systems of thought.

Related questions can be raised in connection to the definition of science. Is it accurate to use contemporary methodological distinctions as opposed to medieval scientists' conceptions of what they were doing? Is it legitimate to treat science and philosophy as distinct entities when these categories were intertwined, as they were in the Greek tradition that informed Islamic science? More important, to what extent can we justify the logical and methodological reduction of specific features of a science into patterns of reasoning and epistemologies, especially in the absence of corroborating articulations in the works of medieval scientists?

Let me illustrate by using the science that is perhaps the most difficult to situate: physics. In the Aristotelian system, physics has two distinct locations. As the science that deals with the physical world, it occupies a lower position than the demonstrative mathematical sciences; but in terms of a larger theory of knowledge—specifically, the derivation of the theoretical principles or foundations of the various sciences (*mabādi'*)—physics is higher than astronomy, many of whose principles derive from physics and metaphysics. So, already in antiquity, there was an ambiguity regarding the exact meaning and location of physics vis-à-vis other sciences. And this ambiguity was accentuated in the Islamic context. A second level of complication derives from our contemporary understanding of physics, which is fundamentally different from Aristotelian physics. Now, building on the argument that the mathematization of nature was the key feature in the emergence of the new sciences, if we try to trace instances in which nature was mathematized in the Islamic (or medieval) practice of science, and if we see, for example, increasing mathematization of astronomy at the expense of natural philosophy, can we justifiably speak of the mathematization of physics?[6] Furthermore, can we make such an assertion if the scientists at the time did not conceive of any such thing as the mathematization of physics or nature, or articulate a mathematical conception of physics? These are questions for which there are no easy answers. Instead of generalizing, let us examine the relationship between science and philosophy through the prism of a single discipline: astronomy.[7]

Astronomy

In *The Exact Sciences in Antiquity*, Otto Neugebauer has the following to say about astronomy: "I do not hesitate to assert that I consider astronomy as the most important force in the development of science since its origin sometime around 500 B.C. to the days of Laplace, Lagrange, and Gauss."[8] Whether we agree with this assessment or not, there can be no doubt that astronomy is one of the oldest and most developed exact sciences, and one of the most esteemed from antiquity on through the scientific revolution.[9] By its very nature, astronomy is an embedded science that partakes of multiple sciences and systems of knowledge: on the one hand, it employs a variety of mathematical disciplines in observation, computation, and construction of geometrical models, and on the other hand, it relates these models to a cosmology in accordance with principles of natural philosophy. Indeed, many of the mathematical sciences were originally developed to facilitate astronomical research. Within the Aristotelian system, even a branch of medicine was developed to explore the effects of heavenly phenomena on health. Various disciplines and belief systems intersect and interact in astronomy, including physics and metaphysics, as well as mathematics and religion. For these and other reasons, the field of astronomy provides fertile ground for identifying the epistemological roots of science and for tracing their historical development. As we shall see, the field of Islamic astronomy was particularly ripe with epistemological discussions pertaining to the relationship between the various systems of knowledge informing astronomy.

Ptolemy's (second century CE) *Almagest* was the main authoritative work that informed the Arabic astronomical tradition.[10] Of all the books of antiquity, the *Almagest* represents the most successful work of mathematical astronomy; its geometric representations of the universe provided the most accurate and best predictive accounts of observed celestial phenomena. A Greek tradition of physical astronomy is reflected both in the *Almagest* and in Ptolemy's other influential work, *The Planetary Hypothesis.*[11] *The Planetary Hypothesis* deals with the physical structure of the heavens and with physical causes underlying Ptolemy's mathematical models of the heavens. Ptolemy believed in the reality of his celestial spheres and adopted, at least in theory, two basic Aristotelian principles: that the earth is stationary at the center of the universe and that the motion of heavenly bodies ought to be represented by a set of uniform circular motions. In practice, mathematical considerations often forced Ptolemy to disregard the second of these principles. Against his better "mathematical" judgment, he professed his adherence to the only physical theory or cosmology available to him: Aristotle's.[12] This was the only way to view the world as an ordered system, and Ptolemy's adherence to this cosmology gave rise in the Islamic period to a long and fruitful tradition of critiquing and reforming Ptolemaic astronomy.

According to the predominantly Aristotelian tradition adopted by Ptolemy, the universe is organized into a set of concentric spheres, each carrying a planet and rotating around the stationary earth at the center of the universe.[13] "Nature," says Aristotle, "is a cause that operates for a purpose."[14] One major distinction between terrestrial and heavenly bodies is their natural principle of motion. Terrestrial bodies are attracted to their natural abode; that is, if left to act according to their nature, all terrestrial bodies will move in linear motion toward their natural place, the stationary center of the earth. In contrast to this sublunary rectilinear motion, the heavenly bodies move in perfectly uniform circles according to their natural principle of motion. Only the four elements (earth, water, air, and fire) of the sublunar region can undergo corruption, which corresponds to their finite linear motion. Heavenly bodies, on the other hand, move in infinite and perfect circular motions, and are made up of an incorruptible fine matter. Each of the seven planets has spheres that, along with the eighth sphere of the fixed stars, make up the finite universe. Within this general Aristotelian cosmology, the motion of the planets is attributed either to a ninth sphere that moves everything that lies below it or to individual souls in the self-moving spheres of each of the planets. In either case, the motion and substance of the heavenly bodies is perfect and incorruptible—and is fundamentally different from the motion and the substance of the sublunar region. This fundamental difference between heaven and earth, as it were, is predicated on the impossibility of vacuum or void.[15]

These cosmological (or natural philosophical or physical) assumptions underlay Ptolemy's astronomy. If Ptolemy had adhered slavishly to these principles, however, he would have had to give up on his profession altogether, since it was not possible to construct accurate mathematical models that could predict the movements of the planets without deviating from—in effect, violating—Aristotelian principles. Astronomers long before Ptolemy realized this. Nonetheless, Ptolemy took for granted the principles of Aristotelian cosmology, including that the heavenly bodies have a divine and eternally unchanging nature. Although his main contribution was in the field of mathematical astronomy, he attributed the failure of his models to conform to Aristotelian principles to human deficiency and asserted that astronomers should strive to make mathematical models that conformed with Aristotelian cosmology.

As we have seen, Arabic astronomy in the ninth and tenth centuries was characterized by a steady move toward systematic mathematization and exhaustiveness. Astronomers examined all aspects of Ptolemaic mathematical astronomy in order to correct its observations and parameters, refine its methods, fill its gaps, strengthen its mathematical foundations, and generalize it in every possible way. By the beginning of the eleventh century, Arabic astronomers had exhausted the possibilities for expanding the frontiers of the field within the limits of Ptolemaic astronomy and were ready to produce a new astronomical synthesis to replace Ptolemy's *Almagest,* one that fulfilled Ptolemy's own research project more comprehensively than he could. At the same time, their command of Ptolemaic astronomy enabled them to make a penetrating assessment of it, which shifted the focus of astronomical research. From the eleventh century onward, the efforts of most theoretical astronomers were directed toward providing a thorough evaluation of the physical and philosophical underpinnings of Ptolemaic astronomy and proposing alternatives to it.[16] Before discussing the epistemological consequences of this tradition of astronomical reform, let me give a brief and simplified overview of some of the problems in the Ptolemaic models and the proposed mathematical solutions of these problems.

In the *Almagest,* Ptolemy used the results of earlier Hellenistic astronomy and incorporated them into one great synthesis. Of particular geometrical utility was the concept of eccentrics and epicycles developed by Hipparchus (second century BCE). In an astronomical representation employing the eccentric model, a planet is carried on the circumference of an eccentric circle that rotates uniformly around its own center *G.* This center, however, does not coincide with the location *O* of an observer on the earth. As a result, the speed of the planet appears to vary with respect to the observer at *O.* In an epicyclic model, the planet *P* is carried on the circumference of an epicycle whose center is in turn carried on a circle called the deferent, which rotates uniformly around the center of the universe, the earth. Viewed by an observer at point *O*, the uniform motions of the deferent and the epicycle produce in combination a non-uniform motion that is mathematically equivalent to the motion of the eccentric model (fig. 1).

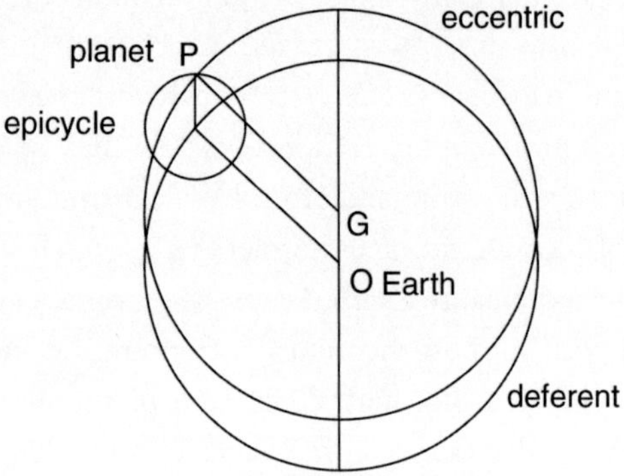

Figure 1

The Ptolemaic model for the motion of the sun utilizes either a simple eccentric or the equivalent combination of a deferent and an epicycle. All the other Ptolemaic models for planetary motions are considerably more complex. For example, in the model for the longitudinal motion of the upper planets Mars, Jupiter, and Saturn, the center *G* of the deferent circle does not coincide with the location of the observer on the earth; moreover, the uniform motion of the center of the epicycle on the circumference of the deferent is measured around the point *E*, called the equant center, rather than around the center *G* of the deferent (fig. 2). Ptolemy proposed this model because it allowed for fairly accurate predictions of planetary positions. However, circle *G* in this model is made to rotate uniformly around the equant *E*, which is not its center. This represents a violation of the Aristotelian principle of uniform circular motion around the earth, the stationary center of the universe. In other words, for

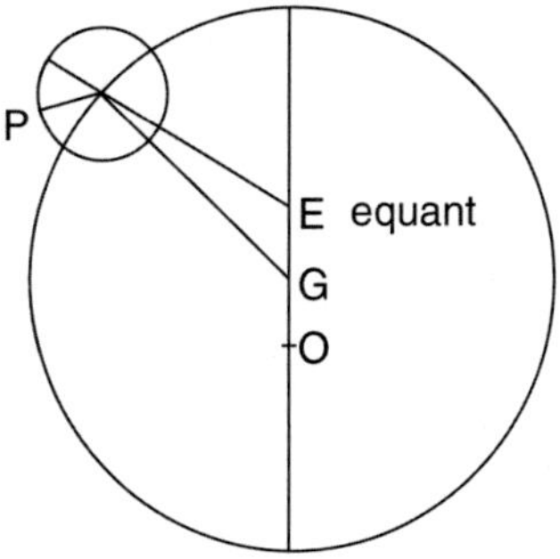

Figure 2

the sake of observation, Ptolemy was forced to breach the physical and philosophical principles on which he built his astronomical theory. Each additional level of complexity in his models brought new objections against Ptolemaic astronomy.

Other objections to Ptolemaic astronomy included the problem of the prosneusis point in the model for the longitudinal motion of the moon, the problem of the inclination and deviation of the spheres of Mercury and Venus, and the problem of planetary distances. Even more problems existed in the case of the moon because Ptolemy's model has a deferent center that itself moves; moreover, the center of the epicycle on this deferent rotates uniformly, not around the deferent's center but around the center of the world. To complicate matters further, the anomalistic motion of the epicycle is measured away from the mean epicyclic apogee that is aligned with a movable point called the prosneusis point, rather than measured from the true apogee, which is aligned with the center of the world. The prosneusis point is the point diametrically opposite the center of the deferent on the other side of the center of the world. The model for the longitudinal motion of Mercury contained complex mechanisms that were equally objectionable. Additional complications resulted from the motion of the planets in latitude: the motion in longitude is measured on the plane of the ecliptic, which is the great circle of the celestial sphere that traces the apparent yearly path of the sun as seen from the earth. The deferents of the Ptolemaic models, however, do not coincide with this plane. Even in the least problematic model, the lunar model, where the deferent has a fixed inclination with respect to the ecliptic, and the epicycle lies in the plane of the deferent, the epicycles of the upper planets do not lie in the plane of the deferent, and they have a variable deviation with respect to it. In the case of the lower planets, both the inclination of the deferent with respect to the ecliptic and the inclination of the epicycle with respect to the deferent are variable. Without going into more details, we can easily imagine the complexity and potential problems of the Ptolemaic models that attempted to account for these seesaw and oscillation motions.[17]

Until now, historians of Islamic astronomy (including myself) have argued that the scholars attempting to solve the problems with the Ptolemaic model fell into two general schools: a mathematically oriented school, which was predominantly in the eastern parts of the Muslim world, and a philosophically oriented school, which was based in the western regions of the empire.[18] Here I will propose a more nuanced classification of the responses to Ptolemaic astronomy, one that takes into consideration fundamental epistemological assumptions. But first, let me say a word about the technical-mathematical aspects of the two types of response.

Astronomers of the eastern reform tradition adopted several mathematical strategies in their attempts to solve the theoretical problems of the Ptolemaic models.[19] One of their main objectives was to come up with models in which the motions of the planets could be generated as a result of combinations of uniform circular motions and still conform to the accurate Ptolemaic observations. Two useful and extremely influential mathematical tools were invented by Al-Ṭūsī and Mu'ayyad al-Dīn al-'Urḍī (d. 1266). The first tool, known in modern scholarship as the Ṭūsī couple, in effect produces linear oscillation as a result of combining two uniform circular motions.[20] The tool was used in various ways by many astronomers, including Copernicus. The 'Urḍī lemma was an equally versatile mathematical tool used by Al-'Urḍī and his successors.[21] To produce optimal representations that are physically and mathematically sound, other astronomers used various combinations of these two tools and devised additional tools. Other mathematical solutions were proposed to resolve the contradictions inherent in the Ptolemaic models. For example, Al-'Urḍī reversed the direction and tripled the magnitude of motion of the inclined sphere in the Ptolemaic lunar model; he was thus able to produce uniform motion around the geometric center of the sphere while at the same time reproducing uniform motion around the old Ptolemaic center. The most comprehensive and successful models were introduced in the fourteenth century by the Damascene astronomer Ibn al-Shāṭir, whose models for all the planets utilize combinations of perfect circular motions where each circle rotates uniformly around its center. Ibn al-Shāṭir was also able to solve problems of planetary distances and to provide more accurate accounts of observations.

The development of Arabic astronomy in Al-Andalus and North Africa followed different routes.[22] The focus of astronomical research in both places in the twelfth century shifted from practical astronomy to planetary theory.[23] The names associated with this research tradition include Ibn Bāja (d. 1138), Jābir Ibn Aflaḥ (fl. 1120), Ibn Ṭufayl (d. 1185), Ibn Rushd (Averroës, d. 1198), and Abū Isḥāq Nūr al-Dīn al-Bitrūjī (fl. 1200). Of these, Al-Bitrūjī was the only one to formulate an alternative to Ptolemaic astronomy; the others produced philosophical discussions about it. Both the discourses on Ptolemaic astronomy and the actual proposed model of Al-Bitrūjī conceived of astronomical reform in reactionary terms—that is, in terms of adopting older and mathematically inferior models in place of the ones used since Ptolemy. The aim of the western school was to reinstate Aristotelian homocentric spheres and to completely eliminate use of eccentrics and epicycles. In accordance with the most stringent and literal interpretations of Aristotelian principles, the western researchers demanded that the heavens be represented exclusively by using nested homocentric spheres and perfect uniform circular motions. Even epicycles and deferents that rotated uniformly around their centers were not tolerated, because their use entailed an attribution of compoundedness to heavenly phenomena; according to Aristotelian principles, the heavens are perfectly simple. Since the predictive power of the Ptolemaic models and their ability to account for the observed phenomena relied on the use of epicycles and eccentrics, the western models were strictly qualitative and philosophical—and were completely useless from a mathematical point of view. They could not be verified numerically, nor could they be used to predict planetary positions. It is no wonder that only one of the western philosophers bothered to produce actual geometrical models.

The epistemological significance of the difference between the eastern and the western reform traditions of Arabic astronomy cannot be overemphasized.[24] Studies of both traditions have focused on their different approaches to the geometric representations of planetary models, but in my view, the more fundamental epistemological differences have yet to be identified. If indeed the purpose of the eastern reform tradition was to ensure that Ptolemaic astronomy conformed to Aristotelian cosmology, then the objectives and the epistemological assumptions of the eastern and western traditions would be similar, and the only difference between them would be over mathematical representation. In fact, the differences between the two approaches were much too fundamental to be reconciled — so how are we to characterize them?

Comparison may be the best way to answer the question. The responses to the epistemological objections raised to Ptolemaic astronomy were more diverse than the eastern-western division suggests. Already in the eleventh century, in place of this division, there were two fundamentally different articulations of the problems with Ptolemaic astronomy, suggesting distinctly different solutions. Within the eastern tradition, as George Saliba has shown, the Arabic astronomical reform tradition culminated in the sixteenth century in the work of Shams al-Dīn al-Khafrī (d. 1550). Certain epistemological assumptions are reflected in the eleventh- and sixteenth-century articulations of this research tradition, which I propose to identify and connect.

By every standard, the eleventh century was a rich intellectual period in Islamic history. Contributions in the realm of the rational sciences were countless, but four individuals stand out as perhaps the most influential thinkers: Abū Rayḥān al-Bīrūnī (d. 1048) and Ibn al-Haytham (d. 1039), in the exact sciences, Ibn Sīnā (d. 1037) in medicine and philosophy, and, a little later, Abū Ḥāmid al-Ghazālī (d. 1111), who provided an influential critique of philosophy.

Several eleventh-century scientists expressed theoretical reservations on aspects of Ptolemaic astronomy. In a book entitled *Tarkīb al-Aflāk*, Abū 'Ubayd al-Jūzjānī (d. c. 1070) indicates that both he and his teacher, Ibn Sīnā, were aware of the equant problem of the Ptolemaic model. Al-Jūzjānī even proposes a solution for the problem. The anonymous author of an Andalusian astronomical manuscript refers to a work that he composed, *Al-Istidrāk ʿalā Baṭlamyūs* (Recapitulation Regarding Ptolemy), and indicates that he included in it a list of objections to Ptolemaic astronomy.[25] But the most important work of this genre was written around the same time by Ibn al-Haytham. In *Al-Shukūk ʿalā Baṭlamyūs* (Doubts on Ptolemy), Ibn al-Haytham sums up the physical and philosophical problems inherent in the Greek astronomical system and provides an inventory of the theoretical inconsistencies in the Ptolemaic models.[26] Many astronomers after the eleventh century took up the theoretical challenge that he outlined, and attempted to rework the Ptolemaic models and to provide alternatives. The list of astronomers working within this tradition comprises some of the greatest and most original Muslim scientists. Those who have received modern scholarly attention include Al-ʿUrḍī, Al-Ṭūsī, Quṭb al-Dīn al-Shīrāzī (d. 1311), Ṣadr al-Sharīʿa al-Bukhārī (d. 1347), Ibn al-Shāṭir, ʿAlāʾ al-Dīn al-Qushjī (d. 1474), and Al-Khafrī.[27]

As I suggested earlier, much of the historical scholarship on this reform tradition has focused on the Ptolemaic planetary models that were deemed problematic and on the alternative models proposed in their place. However, to situate Ibn al-Haytham and the other reformers relative to the Ptolemaic tradition, we need to identify the epistemological grounds of their objections. In other words, what was it that Ibn al-Haytham objected to in Ptolemy's models? This is significant because, as I will try to show, Ibn al-Haytham's objections were fundamentally different from those of Al-Bīrūnī, his contemporary. The tradition that took up Ibn al-Haytham's challenge gradually redefined itself and articulated its criticisms of Ptolemy in a language that was different from Ibn al-Haytham's. So let us take a closer look at the criticisms of Ptolemy.[28]

After listing several observational and mathematical errors

in the work of Ptolemy, Ibn al-Haytham points out Ptolemy's failures to conform to Aristotelian cosmology. In reference to his lunar model, for example, Ibn al-Haytham says that this model attributes two contradictory natural motions to the single solid orb of the epicycle. This, Ibn al-Haytham adds, is "an utter impossibility." And if these two motions are voluntary, then "a part of the heavens" must make "two opposing choices, which means that its substance is composite and made out of two or more contradictory essences, which is an *impossibility, according to all philosophers*" (19; my emphasis). Later, in the models for the wandering planets, Ibn al-Haytham criticizes Ptolemy's principles of the eccentric and the epicycle, arguing that the only acceptable principles are ones that entail circular and uniform motions (23 ff.). Moreover, Ibn al-Haytham recurrently maintains that perceptible motions can be ascribed only to existing real solids; and if that is true, no solid in the sublunar world of generation and corruption will have at once two natural and opposite motions (34 ff.). This is more the case for heavenly bodies, because they are all made of one essence, which admits of no contradiction. Ibn al-Haytham says that all competent philosophers believe that there can be no two opposing motions in the heavens, especially if the opposing motions are ascribed to a single heavenly solid (37).

In his criticism of Ptolemaic models, Ibn al-Haytham's frame of reference is clearly philosophical. He clarifies his stand further when he insists that part of the problem in Ptolemy's astronomy is that he performed his calculations on imagined circles and lines, not on real solids, whereas the motions of real planets cannot be validly described by configurations (*hay'a*s) that cannot have a real existence (38). Ptolemy's main failure, then, lay in his inability to propose a configuration that is mathematically accurate and has a physical reality. According to Ibn al-Haytham, to imagine circles in the sky and then imagine that planets move in those circles is not sufficient to make planets actually move in those imagined circles. Ibn al-Haytham concludes his critique of the *Almagest* by saying that the motions of the planets have correct configurations or models (*hay'a ṣaḥīḥa*) in real physical solids, which Ptolemy was not able to find (40–42, 64).

In critiquing the *Planetary Hypothesis*, Ibn al-Haytham maintains that Ptolemy attempted to provide solid models for the motions of the planets that could have real existence in the heavens and that would have the same nature as the heavenly bodies. The problem, however, as Ibn al-Haytham sees it, is that the configurations (hay'as) of the *Almagest* do not produce the same results as the solid bodies in the *Planetary Hypothesis*. Another objection is to Ptolemy's sectional (*manshūrāt*) solid models themselves, which, according to Ibn al-Haytham, create two main impossibilities: they require contradictory movements of the individual heavenly solids, which is impossible for solids made of one essence, and they require that a heavenly body vacates one location and fills another, which necessitates the existence of vacuum or void, another impossibility in Aristotelian cosmology (46, 50, 59).[29]

In criticizing Ptolemy for his failure to produce mathematical models that can have real physical existence, Ibn al-Haytham repeatedly refers to principles determined by the philosophers—more specifically, those principles determined by analogy (*qiyās*) from the sciences of metaphysics and physics (*ṭabī'ī*).[30] Therefore, the primary criteria used in assessing the adequacy of the Ptolemaic models are from Aristotelian natural philosophy, which purports to describe and explain the real physical existence of the heavenly bodies, not just the motions traced by these bodies. To be sure, Ptolemy also subscribed to the same natural philosophy and worked within the contours of an Aristotelian cosmology, but Ibn al-Haytham points out the instances in which Ptolemy failed to keep to his physics. Of course, Ibn al-Haytham might have been a little more forgiving of Ptolemy had he attempted to construct the alternative models that he himself demanded.

Be that as it may, the issue here is not who scores more points. Rather, I am trying to underscore two points. First, Ibn al-Haytham's critique was totally within the framework of Aristotelian natural philosophy, which puts him squarely within the tradition of Ptolemy *and* Aristotle. Second, although Ibn al-Haytham did not depart epistemologically from Aristotle, whose natural philosophy provided the basis for the knowledge about heavenly bodies and therefore the demonstrable principles of astronomy, the importance of Ibn al-Haytham's *Al-Shukūk* was that it articulated a perception of crisis within the discipline of theoretical astronomy, a crisis that no serious astronomer after Ibn al-Haytham could ignore. The history of Islamic theoretical astronomy from the eleventh to the sixteenth century is the history of the attempts to reconceptualize and resolve this crisis.[31]

Ibn al-Haytham was not alone in suggesting problematic aspects of Ptolemaic astronomy that needed to be addressed and corrected. In fact, a clearer, though much shorter expression of the same sort of objection that Ibn al-Haytham aired was expressed by Ibn Sīnā, who was primarily a philosopher and a physician; he was not as accomplished in the mathematical sciences as Ibn al-Haytham. In the volume dedicated to astronomy in his philosophical encyclopedia, *Al-Shifāʾ*, Ibn Sīnā describes the task for astronomers as follows: "We need to correlate what is mentioned in the *Almagest* with what is reasoned from natural science (*al-maʿqūl min al-ṭabīʿī*) to know how these movements occur and to incorporate [in the planetary models] the benefits that were reached after the time of the *Almagest* but still correspond to the positions [observed] in the *Almagest*. The first thing that is needed is to figure out the manner in which it is possible for a sphere to rotate around itself while it is inside another sphere and while it follows the encompassing sphere in its motion." I am suggesting here that this statement by Ibn Sīnā is clearer than the more elaborate descriptions by Ibn al-Haytham because Ibn Sīnā directly relates the tasks he sets for astronomers to the scientific traditions in question. To start with, the proposed tasks fall within the parameters of the Ptolemaic project; the astronomer needs to account for Ptolemaic observations as well as for corrections and discoveries accumulated in the Arabic astronomical tradition. At the same time, however, the models of the *Almagest* must also conform to the principles extracted through reasoning, presumably deductive reasoning, from al-'ilm al-ṭabī'ī—that is, from physics, or, more accurately, from natural philosophy.[32]

Yet if this reformative approach to Ptolemaic astronomy was conceptualized in the context of Aristotelian physics, another critical tradition was aimed directly at the Aristotelian natural philosophy that informed Ptolemy's cosmology. This criticism was articulated by Al-Bīrūnī.

All three thinkers, Ibn al-Haytham, Ibn Sīnā, and Al-Bīrūnī, were brilliant in their respective fields. Ibn al-Haytham was a prolific scientist who composed treatises in the various mathematical sciences, including some on applied astronomy; he also composed some philosophical works, but the main contribution for which he is known is *Kitāb al-Manāẓir*, a book that revolutionized the science of optics. Ibn Sīnā was the greatest and most influential Aristotelian philosopher in Islamic history. He was also a distinguished physician whose book *Al-Qānūn fī al-Ṭibb* was probably the single most influential work of medicine in the whole world until the seventeenth century. His philosophical encyclopedia, *Al-Shifā'*, included books that dealt with the mathematical sciences, but his contribution in these fields was not of the same caliber as his contributions in philosophy and medicine. Al-Bīrūnī, who was also extremely prolific, composed works on pharmacology and metallurgy and a variety of other fields, including an anthropological history of India and even some minor philosophical works. The vast majority of his scholarship, however, was in the field of astronomy, and he was certainly the most accomplished astronomer of the Arabic tradition up to the eleventh century and one of the greatest scientists in Islamic history. I mention these credentials to suggest that the views we are examining here are at the heart of the Arabo-Islamic scientific and philosophical traditions and not the inconsequential views and guesses of individuals on the margins. But I also mention these credentials to situate the critiques by each of these thinkers within the intellectual context of their overall thought. Al-Bīrūnī's critique of Aristotelian cosmology is notable because he was by far the most accomplished astronomer of the three, which suggests that his critique might provide an accurate representation of the professional Arabic stand on Greek astronomy. So what does Al-Bīrūnī have to say?

While Ibn al-Haytham criticizes Ptolemy's failures to conform to Aristotelian cosmology, Al-Bīrūnī reverses the order of criticism and directs it instead at Aristotelian cosmology, which, he argues, the mathematical astronomer is not bound to employ. Al-Bīrūnī conveniently directs his criticism of Aristotle's *De Caelo* (On the Heavens) to Ibn Sīnā, the most celebrated Muslim philosopher of all time. In a book entitled *Al-As'ila wa'l-Ajwiba* (Questions and Answers), Al-Bīrūnī presents Ibn Sīnā with a set of questions in which he criticizes Aristotle's physical theory, especially as it pertains to astronomy. Ibn Sīnā responds, and a lively debate ensues.[33]

Let me preface my discussion of this book by referring to a statement that Al-Bīrūnī makes in another treatise, *Istīʿāb al-Wujūh al-Mumkina fī Ṣanʿat al-Asṭurlāb* (Comprehending All the Possible Aspects of the Craft of the Astrolabe). In this work on the theory and construction of astrolabes, Al-Bīrūnī refers to "a kind of simple astrolabe, constructed by Abū Saʿīd al-Sizjī . . . known as the boatlike astrolabe. I liked it immensely, for he invented it on the basis of an independent principle that is extracted from what some people hold to be true—namely, that the absolute eastward visible motion is that of the earth rather than that of the celestial sphere. I swear that this is an uncertainty that is difficult to analyze or resolve. The geometricians and astronomers who rely on lines and planes have no way of contradicting this [theory]. However, their craft will not be compromised, irrespective of whether the resulting motion is assigned to the earth or to the heavens. If it is at all possible to contradict this belief [in the motion of the celestial sphere] or to resolve this uncertainty, then such [a task] will have to be assigned to the natural philosophers."[34]

Al-Bīrūnī maintains that it does not matter whether the astronomer uses a geocentric or a heliocentric model as far as mathematical astronomy is concerned. This is so because the relative motions will be the same, and the difference amounts to a simple transfer of coordinates. The observational technology available to mathematical astronomy at the time of Al-Bīrūnī (and, indeed, Copernicus) was not accurate enough to provide a satisfactory answer to the question of which model to use. Thus, Al-Bīrūnī concludes that the discussion pertaining to the nature of the motion of heavenly bodies is primarily philosophical, not mathematical, and since he thought of himself as a mathematical astronomer, he did not feel that it was his responsibility to address this philosophical question. Al-Bīrūnī here makes a distinction between mathematical astronomy and philosophy not just on the basis of professional affiliation but also on epistemological grounds, that is, on the basis of the kinds of knowledge relevant to each discipline.[35] As many historians of science have noted, what was lacking until Kepler and Newton was a physical theory to replace the old Aristotelian system of natural philosophy, and the occasional alternatives to aspects of this philosophy certainly did not cohere into a full system. In effect, Al-Bīrūnī recognizes that he has no grounds for positing an alternative to Aristotelian cosmology on the basis of mathematical astronomy alone, but he also suggests that his discipline is not bound by the prevalent principles of natural philosophy.

Al-Bīrūnī's questions and Ibn Sīnā's equally informative answers amount to a systematic discussion of the nature of the relationship between science and philosophy. In the questions, Al-Bīrūnī makes a number of assertions that are incompatible with some of the most fundamental elements of Aristotelian natural philosophy—and the logical structure of Aristotelian natural philosophy, which was further systematized in the Arabic philosophical tradition, meant that a hole in the system could undermine it in its totality. In fact, Ibn Sīnā's responses to Al-Bīrūnī illustrate the interconnectedness of various aspects of the philosophy. He draws on all of Aristototle's books in his responses, not just *De Caelo*, the only book questioned by Al-Bīrūnī.[36] It is clear that in Ibn Sīnā's mind, undermining some of the arguments of *De Caelo* would bring into question the whole Aristotelian corpus of natural philosophy.

Al-Bīrūnī commences his critique by questioning the central Aristotelian concept that the heavenly bodies admit neither levity nor heaviness and the equally central Aristotelian denial of the possibility of void or vacuum outside the heavenly sphere.[37] Significantly, Al-Bīrūnī does not argue that the heavenly sphere is either heavy or light, nor that there is void outside it, but simply that these possibilities cannot be ruled out. Al-Bīrūnī targets yet another principle of Aristotelian cosmology, that the circular motion of the heavenly orb is natural to it and cannot be accidental or compulsory.[38] He also asserts, against Aristotle, the possibility of other worlds, either ones that have different natures or ones that have the same nature as our world but different principles of motion.[39] Ibn Sīnā argues that multiple worlds are impossible.

In *Al-As'ila*, the debate between Al-Bīrūnī and Ibn Sīnā hinges largely on the kinds of proof each accepts. For example, in criticizing Aristotle's denial of indivisible atoms, Al-Bīrūnī concedes that positing indivisible atoms leads to problematic conclusions that the geometricians are familiar with, but he adds that the logical consequences of Aristotle's denial are even worse (17–18). In another example, Al-Bīrūnī undermines an argument that Aristotle uses to posit the impossibility of egg-shaped or lentil-shaped (elliptical) motion by the heavenly bodies, even though, Al-Bīrūnī adds, he does not disagree with Aristotle's conclusions. Aristotle argues in *De Caelo* that the assumption of these shapes necessitates that heavenly bodies vacate one location as they move to another, which would necessitate the existence of vacuum, an utter impossibility in Aristotelian physics. Interestingly, Al-Bīrūnī, who does not concede the impossibility of vacuum, does not suggest that the possibility of vacuum implies the possibility of lentil- or egg-shaped motion. Rather, Al-Bīrūnī undermines Aristotle's argument on mathematical grounds. "The egg shape," Al-Bīrūnī argues, "results from the rotation of an ellipse around its major axis, and the lentil shape results from its rotation around its minor axis. . . . Now if the egg shape rotates around its major axis, and if the lentil shape rotates around its minor axis, then their rotations would be similar to that of a sphere, and no void would be generated because of their rotations. That would only result if the axis of rotation of the egg shape were its minor axis, and for the lentil shape, its major axis, in which case [vacuum] would be necessitated." Al-Bīrūnī concludes: "I do not say this because I think the celestial orb is not spherical, or because I think that it is egg-shaped or lentil-shaped; in fact, I have tried to disprove this [elsewhere]; rather, I am objecting to the logic of [Aristotle's] argument" (27–28). What is interesting here is that although Al-Bīrūnī does not agree with the Aristotelian argument that vacuum is impossible, he does not use the contrary argument in his critique of Aristotle.[40] Rather, he uses mathematical arguments from within his disciplines—geometry and mathematical astronomy. While questioning the natural philosophical arguments of Aristotle, Al-Bīrūnī does not propose an alternative natural philosophy. His aim is to undermine the claims of natural philosophy over his discipline, mathematical astronomy, not to deny or assert the possibility of natural philosophical knowledge outside his discipline.[41] As in his reference to the geocentric universe of the boatlike astrolabe, which raises a question for philosophers, not astronomers, Al-Bīrūnī is willing to grant natural philosophers control over their field, but these philosophers no longer provide the principles, axioms, and postulates of the science of astronomy.

At times, even Ibn Sīnā seems to allow a demarcation between disciplines. For example, in his response to the above argument by Al-Bīrūnī, Ibn Sīnā says that "one can disprove the possibility of an egg-shaped or lentil-shaped orb by using either natural [philosophical] proofs or mathematical, geometric proofs; had I not known of your erudition in the mathematical subjects, and the high level of knowledge of geometry among scholars in your region, I would have provided some [mathematical proofs]" (29). In response to a question by Al-Bīrūnī on the reasons for restricting habitation to one quarter of the earth, to the exclusion of the other northerly quarter and the two southerly quarters, Ibn Sīnā says that certain regions are uninhabitable because of either excessive heat or excessive cold. And "this is as far as my profession (*ṣīnā'atī*) is concerned. As for finding the quantities [i.e., coordinates] of a location that does not have causes that prevent it from being inhabited, this is the specialty of the experts in the mathematical sciences; had I not known of your command of this field, I would have referred to aspects of the geometrical science" (41–42).

Although Ibn Sīnā systematically uses knowledge from one subfield of the philosophy to prove arguments in another, he seems to concede in his debate with Al-Bīrūnī a degree of epistemological specialization, not just professional distinctions between fields. Of course, Al-Bīrūnī goes much further than Ibn Sīnā by suggesting that his discipline of mathematical astronomy can be insulated from arguments in natural philosophy and that its principles are fully internal to itself. And once again, the distinctions that Al-Bīrūnī draws between a mathematical astronomer and a philosopher are epistemological and not just professional or vocational. The natural philosophical principles (including physics and metaphysics) on which philosophers build their physical theories do not constitute valid evidence for the mathematical astronomer. On the other hand, while recognizing Al-Bīrūnī's mathematical skills, Ibn Sīnā tends to dismiss Al-Bīrūnī's ability to engage in larger philosophical debates. In one instance, Al-Bīrūnī marvels at the weakness of one of Aristotle's arguments about the unchanging nature of the heavenly bodies, which is based on what bygone generations said. Al-Bīrūnī says that there is very little valid evidence on any subject that has reached us from the past. In responding, Ibn Sīnā does not try to defend Aristotle's argument; rather, he dismissively says to Al-Bīrūnī that he must "have taken this objection from John the grammarian, who wanted to mislead Christians by pretending that he disagrees with Aristotle on this matter, although if you look at the end of his commentary on [Aristotle's] *Generation* and *Corruption* and other books, you will not fail to note his agreement with Aristotle on this matter. Or, you could have taken this critique from Muḥammad Ibn Zakariyyā al-Rāzī, whose pretentiousness in meddling with metaphysics made him overestimate his abilities, [which were limited to] dressing wounds and testing urine and feces" (12–13).[42]

The point is not who wins the debate, or who is smarter. The point is that fault lines were being drawn between philosophy and science (astronomy, in this case) in the debate itself. To reiterate, *Al-As'ila* was not an attempt to provide a general theory of knowledge that defines and demarcates within it the purview of the science of astronomy in relation to other possible forms of knowledge. Rather, Al-Bīrūnī, who deliberately and consciously restricted his positive pronouncements to his own field, suggests that the conditions for astronomical knowledge are determined within the field of astronomy itself and by using the mathematical tools appropriate to it. Al-Bīrūnī was an astronomer, not a cosmologist, and his interest was in the quantitative description of the motions of the planets, not in the nature of celestial bodies and the causes of their motions. This makes Al-Bīrūnī's questions all the more remarkable, since they were critiques of the application of natural philosophy, of physics and metaphysics, to astronomy; what gives Al-Bīrūnī's questions added weight is that they came from a mathematical astronomer, not from a philosopher. In contrast to Ptolemy, who noted the discrepancy between his models and those of Aristotelian natural philosophy but continued to subscribe to the latter and explained away the former as a failure on his part, Al-Bīrūnī refused to concede the intellectual authority of systems of knowledge outside his own system. By questioning the demonstrability of the principles of physics, Al-Bīrūnī was asserting the right of astronomy to derive its own principles within its disciplinary domain. This is the deep meaning of Al-Bīrūnī's refusal to assign a nature to the heavenly bodies and his equally radical attempt to undermine the absolute Aristotelian distinction between the celestial and terrestrial realms.[43] In Al-Bīrūnī's discipline, the motions of the heavenly bodies can be subjected to the same observational criteria that apply to the earth; both realms can be described and even explained by the same tools and by using the same principles.

I do not mean to suggest that the *Al-As'ila* was an articulation of an alternative cosmology that provided the exclusive research agenda for astronomers after Al-Bīrūnī. In the short run, Ibn al-Haytham's *Al-Shukūk* had a larger impact than Al-Bīrūnī's *Al-As'ila*, partly because Ibn al-Haytham articulated a program of research and presented astronomers with problems to solve. Al-Bīrūnī's critique, on the other hand, was aimed at dissociating astronomy from natural philosophy but without proposing specific new tasks for astronomers. In the long run, the implications of Al-Bīrūnī's critique were much more significant than those of Ibn al-Haytham's. The initial attempts to reconcile Ptolemaic mathematical astronomy with Aristotelian natural philosophy were gradually diluted, and of the various sciences that informed astronomy, the ones that endured were mathematical. Ultimately, therefore, the theoretical and conceptual commitments of the astronomers shifted, and the view that prevailed was Al-Bīrūnī's, because it resonated with other cultural forces outside the sciences.

Let us fast-forward to the sixteenth-century astronomer (also ḥadīth scholar) Shams al-Dīn al-Khafrī, whose approach to planetary theory is documented by George Saliba.[44] Al-Khafrī presents thorough accounts for the various alternative models proposed by earlier astronomers and adds a few of his own invention. Saliba shows that the purpose of Al-Khafrī's presentation was not to look for a correct model, nor to decide which model conforms with an ideal or preferred cosmology, but to establish the mathematical equivalence of all of these models. In Al-Khafrī's production of several models to represent the same observational data Saliba sees what he calls "a new departure," one "in which mathematical models were perceived to be as different ways (*wujūh*) of saying the same thing. Mathematics, then, as a tool for the astronomer, is just like a language. One can use it to express himself in different manners." Saliba notes that with Al-Khafrī, "there is no question of cosmology, in the sense posed by Ptolemy. In all of Khafrī's models, both Ptolemaic hypotheses [eccentrics and epicycles] are used, and yet Khafrī finds that he can offer many solutions to each of the problems that he encounters."[45] "This attitude,' says Saliba, "is obviously new with Khafrī."[46]

Al-Khafrī did not privilege any specific model, because all models were mathematically equivalent. As far as he was concerned, they were interchangeable, nor did he seek to identify the one that conformed most with Aristotelian cosmology.[47] What was conspicuously absent in Al-Khafrī's work was any attempt to fit nature, or what was conceived of as nature in Aristotelian natural philosophy, into any particular model. The organizing principle for Al-Khafrī's work was mathematical equivalency, which implies that no one correct model is to be singled out on account of its correspondence with reality. Of course, there is no reason to think that Al-Khafrī believed in multiple realities. Rather, the object of his study was the mathematical models constructed by astronomers and not the outside physical reality that the models purportedly represented. Mathematical representations, in this sense, were no longer a mere medium for translation between reality and observation; they were an object of study in themselves.

In another book relevant to our theme, Jamil Ragep focuses on an aspect of the late medieval debate among Muslim astronomers on the role of Aristotelian physics in astronomy: the question of the rotation of the earth.[48] Among the diverse views expressed on this matter, Ragep focuses on that of 'Alā' al-Dīn al-Qushjī, who forcefully and unambiguously argued that the science of astronomy "does not depend upon physical (*ṭabī'iyya*) and theological (*ilāhiyya*) premises (*muqaddimāt*)."[49] In keeping with this principle, Al-Qushjī dispensed with natural philosophy altogether, both in his theoretical argument and in the organization of his mathematical work. Although Al-Qushjī rejected the idea, based on observational criteria, that the earth moves, he nonetheless asserted that assuming that the earth moves would not result in anything false in natural philosophy. As with Saliba's assessment of Al-Khafrī, Ragep suggests that Al-Qushjī's rejection of natural philosophy made him "almost unique among medieval astronomers and philosophers."[50]

Yet if Al-Qushjī and Al-Khafrī were unique in their attitudes toward the role of metaphysics in astronomy, how can we explain their corresponding approaches and the similarity of their attitudes to Al-Bīrūnī's in the eleventh century? To support the view that Al-Khafrī was taking a new departure, Saliba compares him to Al-Shīrāzī, another of the great and original Muslim astronomers, who "proposed nine models to solve the problem of Mercury's equant" but concluded that only one was correct. In contrast, Saliba notes, Al-Khafrī added several models of his own and of Al-Qushjī but introduced none as correct.[51] Likewise, to underscore Al-Qushjī's distinctiveness, Ragep lists the views of some of the major astronomers, most notably Al-Ṭūsī and Al-Shīrāzī, and notes: "As astronomers, Ṭūsī and his successors needed, indeed demanded, some conclusive proof concerning a matter of such basic importance to astronomy [as the motion of the earth]."[52] Al-Ṭūsī, in most of his astronomical work, tries to establish the science of astronomy on mathematical principles, but he also maintains that occasionally the astronomer needs to use the results of the natural philosophers. The principles of astronomy that "need proof," he says, "are demonstrated in three sciences: metaphysics, geometry, and natural philosophy."[53] For example, Al-Ṭūsī goes over the observational arguments usually used to prove that the earth is stationary, only to conclude that the proof has to be sought in natural philosophy: "It is on account of the [earth] having a principle of rectilinear inclination that it is precluded from moving naturally with a circular motion."[54] Al-Shīrāzī, on the other hand, disagrees with Al-Ṭūsī and maintains that observation can determine that the earth is stationary. He also insists on the "need to establish the science of astronomy (i.e., *'ilm al-hay'a*) without recourse to natural philosophy" and instead to base it on "observation and testing (*al-raṣd wa'l-i'tibār*)."[55] Clear as this statement is, Al-Shīrāzī still included in his astronomical works sections on natural philosophy, a fact that Ragep takes as an indication that Al-Shīrāzī did not advocate a full break with natural philosophy, as Al-Qushjī did.

Without underestimating the originality of Al-Qushjī or Al-Khafrī, or, for that matter, most of the astronomers attempting to reform Ptolemaic astronomy, I would suggest that we can still look at the work of Al-Qushjī and Al-Kahfrī as culminations of a trend rooted in earlier astronomical practice. As already noted, the questioning of the Aristotelian cosmology that underlay Ptolemy's astronomy occurred within the science of astronomy itself and outside it in direct dialogue with natural philosophy.

That many, though not all, astronomers in the thirteenth and fourteenth centuries continued to discuss natural philosophical issues in their astronomical works does not exclude the possibility of a conceptual shift in their treatment of natural astronomy. Equally important is that natural philosophical questions were usually treated as observational testing grounds, even in the few cases where they were invoked. The question of the motion of the earth, for example, was often addressed by considering such observational questions as, What would happen to an object thrown vertically upward? and How do we explain the motions of clouds and comets?[56] Although Al-Qushjī articulated an argument for excluding natural philosophy from astronomy, and although Al-Khafrī did not feel a need to justify his total disregard of natural philosophy and his exclusive focus on mathematical modeling, most other astronomers who were not as direct and clear in their position did in fact dodge the question of the reality of their mathematical models and focused predominantly on mathematics. Even Al-Khafrī could not have thought or claimed that all the models he listed corresponded to physical reality. Nonetheless, he did assert that his models were mathematically correct. In other words, Al-Khafrī was not denying the existence of a physical reality; he was simply not interested in pursuing it. Al-Shīrāzī, too, described nine mathematically correct models, although unlike Al-Khafrī, he chose a correct one on the basis of its correspondence to reality. There is not much difference between Al-Shīrāzī and Al-Khafrī, however, for the primary focus of their work was mathematical models. The former paid lip service to natural philosophy, and the latter turned his back on it.

The point I am trying to make here is that almost all astronomers dedicated most of their actual work to mathematical astronomy; the attention they paid to cosmology and natural philosophy was minimal, irrespective of what they thought of these fields. The actual practice of astronomy was, therefore, quite removed from natural philosophy and was mostly concerned with mathematical tools and models that could account for actual observations. Occasionally, astronomers gave preference to a particular model on account of its correspondence to reality, yet the bulk of astronomical research was largely mathematical, and it was the prevalence of mathematical practice in astronomy that gave Al-Khafrī's mathematical equivalency approach its epistemological sanction.

Al-Shīrāzī did not assert a definite correlation between his principles and the real universe, although he required the possibility (*jawāz*) of correlating the observed heavenly motions with the proposed models. In other words, although Al-Shīrāzī recognized that several possible configurations could represent the motions of the planets, he chose the one that was logically possible. This does not mean that Al-Shīrāzī was positing an identity between model and reality, nor did he assert a natural certainty for the model he chose.[57] In fact, in his *Tuḥfa*, Al-Shīrāzī introduces four *uṣūl* (hypotheses or principles) that underlie his models.[58] These principles are the eccentric and the epicycle plus the *aṣl al-muḥīta* (epicyclet) and the *aṣl al-kabīra wal-ṣaghīra* (large and small circles). The first two are staples of Greek astronomy. The epicyclet, a small additional epicycle carried on the rim of the first epicycle, was introduced to correct Ptolemaic models. The large and small circles hypothesis is the Ṭūsī couple, made up of a small circle rotating inside a large circle so that the combined motions of the two circles produce an oscillating rectilinear motion.[59]

These mathematical tools had great utility and were widely used in the reformed astronomy, but what I would like to underscore here is Al-Shīrāzī's peculiar use of the term *aṣl* (plural, *uṣūl*). His hypotheses were merely mathematical tools, not akin in any way to the Aristotelian natural philosophical principles to which Ptolemaic astronomy was bound. For most astronomers, these were the kinds of principles that mattered most, and Al-Shīrāzī was not an exception. As we have seen, Ibn al-Haytham's *Al-Shukūk 'alā Baṭlamyūs* provided later theoretical astronomers with an articulated set of problems with Ptolemy's model to try to solve. For him, the main problem was the discrepancy between the mathematical models and the real world, between mathematical astronomy and natural philosophy. Numerous astronomers tried to resolve this problem and to develop models that conformed to the principles of this astronomy. But as astronomers soon realized, the quest for perfect harmony between mathematical astronomy and natural philosophy, if taken literally, will be futile.

The serious attempts to achieve perfect harmony were made by the astronomers of the Muslim west, and they were all, without exception, total failures. For the sake of perspective, let me say a few words about the epistemological assumptions of this relatively inconsequential tradition. Most contributors to it were better known as philosophers, and in most cases they did not produce more than short descriptive accounts of the natural philosophical conditions that a valid astronomy must fulfill. The only acceptable models, according to this tradition, were concentric spheres with neither eccentrics nor epicycles. According to Mūsā Ibn Maymūn (Maimonides, d. 1204), Ibn Bāja rejected epicycles because they required a rotation around a point other than the center of the earth. Ibn Maymūn says that he has heard indirectly from Ibn Bāja's students that Ibn Bāja constructed models that made no use of epicycles and used only eccentrics. If true, adds Ibn Maymūn, the models had little use, because even eccentrics are a violation of the principle established by Aristotle.[60] We know that one astronomer, Al-Biṭrūjī, did try to construct some models.[61] And Al-Biṭrūjī refers to an attempt by Ibn Ṭufayl to construct models in which no epicycles or eccentrics were used.[62] Ibn Rushd stipulates that the correct models should only posit "configurations that stand for causes (*tajrī majrā al-asbāb*)" and that do not result in "impossibilities in the natural science (*muḥal fī al-ʿilm al-ṭabīʿī*)." He clarifies what he means when he says that movements that appear through observation, such as retrograde motion or the slowing and speeding of planets, are impossible given the "nature of the motion of celestial bodies" as derived from natural science.[63] In other words, in this strict order of hierarchy mathematical astronomy is not allowed to inform or modify the kind of (philosophical) astronomy that describes and explains the real heavenly bodies. Ibn Rushd maintains that "nothing in the science of astronomy (hay'a) of our times deals with real existence; rather, the science of astronomy in our times corresponds to mathematical computation (*ḥisbān*) and not to reality."[64] Elsewhere, Ibn Rushd asserts that "a mathematical celestial body is [conceptually] inferior to a natural celestial body."[65] He also maintains that the only acceptable orbs are ones that have planets in them, which he calls planeted orbs (*falak mukawkab*); on this basis, he rejects the ninth orb, which accounts for the slow motion of the outer celestial sphere of the stars, because every motion of a planet results from "a desire specific to it, and a specific desire (*tashawwuq* or *shawq*) is due only to a specific object of desire."[66] Ibn Rushd then posits an organic universe with no need for centers and epicycles: "It is most fitting to imagine that the whole celestial orb in its totality is one spherical animal and that its various parts move like the organs of an animal. This is why these [partial individual] motions do not need centers to move around, unlike the earth, which is [needed] for the great motion."[67]

The importance of this western Islamic counterrevolution[68] does not lie in any lasting effects it had in the field of astronomy, because there were none.[69] Rather, the counterrevolution shows what the eastern Islamic reform tradition was not. Even Ibn al-Haytham's call to reconcile astronomy, a mathematical discipline, with natural philosophy (al-'ilm al-ṭabī'ī) was unacceptable to someone like Ibn Rushd, who demanded subjugation and not reconciliation. In a discussion of the rainbow, for example, Ibn Rushd objects to Ibn al-Haytham's mixing of natural science and mathematical optics, although mixing is exactly what enabled Ibn al-Haytham to make his optical breakthrough. Natural science, according to Ibn Rushd, provides the general overall principles and leaves the details of specific causal relations to the lower disciplines, presumably like optics.[70] In fact, there can be no demonstrative proof (burhān) that contradicts the general and certain principles that apply to all beings (*mawjūdāt*); otherwise, these would not be general principles.[71]

This western Islamic philosophical project that peaked with Ibn Rushd had a general epistemological agenda that applied to all the sciences, including astronomy. The kernel of this project was to reformulate Aristotelian philosophy by extracting it from the dialectical (*jadalī*) context in which it was originally formulated and restructuring or reformulating it within a demonstrative (*burhānī*) context that Ibn Rushd believed to be pure Aristotelianism.[72] This is why Ibn Rushd composed many different summaries and commentaries on the works of Aristotle, some of which preserved the Aristotelian dialectical style and others of which presented Aristotelianism as a demonstrative system of thought.[73] In his quest to restore "pure" Aristotelianism and to reinstitute demonstrative proof (burhān) as the logical foundation of philosophy, Ibn Rushd did not take issue just with scientists or Muslim critics of philosophy; he also blamed Ibn Sīnā and Abū Naṣr al-Fārābī (d. 950) for introducing many erroneous concepts that were foreign to Aristotelian philosophy.[74] In one instance, Ibn Rushd criticizes Ibn Sīnā's argument that a heavenly solid is composite and made up of form (*ṣūra*) and prime matter (*hayūlī*); Ibn Rushd says that it is a simple solid and admits no compoundedness.[75]

In contrast to Ibn Rushd, both Al-Fārābī and Ibn Sīnā seem to have recognized the difficulty of extracting dialectics (*jadal*) from demonstrative science (burhān). For them, jadal seems to have two functions: on the one hand, burhān depends on abstraction from results reached through jadal, and on the other hand, jadal serves as a tool for testing knowledge. In Al-Fārābī's and Ibn Sīnā's understanding of Aristotelian philosophy, then, jadal and burhān are intertwined. Jadal is not just an inferior form of knowledge between the higher demonstrative knowledge of the philosophers and the lower knowledge (or lack thereof) of the commoners; it is part of that higher knowledge. That said, for Muslim philosophers in general, knowledge was in a strict hierarchy, with the highest form being knowledge of the cause that necessitates the existence of something, as opposed to knowledge of accidents that occur to it. Ibn Rushd insisted on identifying the universal principles (*kulliyāt*) of the sciences, which are derived or proved outside the various partial disciplines and nonetheless bind these disciplines together. Ibn Rushd's insistence was perhaps an exaggerated reaction to the dominant trend in the practice of science and, in his view, even in the practice of philosophy: to submit various disciplines to principles derived from within, thereby threatening the epistemological unity of the Aristotelian system that he sought to redeem.

Medicine

Another example from the field of medicine will illustrate the difference between Ibn Rushd's and Ibn Sīnā's adherence to Aristotelian philosophy and how this difference impacted the science they produced.[76] Islamic medical writers adopted the ancient idea of inverse similarity between male and female sex organs, whereby the cervix of the uterus is the equivalent of the penis. While the male sexual organ is complete and extends outward, however, a woman's organ is "incomplete and held back on the inside (*muḥtabisa fī al-dākhil*), as if it were an inverted male organ." Just like men, women have two testicles, but they are smaller and are embedded inside the vagina.[77]

These types of comparisons between male and female sexual organs reflect the tension in Islamic medical discourse between the idea of equivalent inverse sexualities and the assertion that female attributes are all smaller and fewer than their male equivalents. This tension is evident in the various theories that deal with the role of female sperm in conception and women's agency in the reproductive process. Aristotelian physiology admits of no female testicles or female sperm. Having denied the existence of female semen, Aristotle maintained that woman's contribution to the body of the fetus was the menstrual blood, which was the matter that the semen of the man acted on. A man's contribution, therefore, was in giving both form and the principle of movement, whereas a woman's contribution was in giving passive matter. Hippocrates, in contrast, maintained that the sperm originated in the brain; Galen, that it originated in the liver. For both, the heat of lust and sexual excitation caused all parts of the body to release particles that formed semen. Galen found further support for the Hippocratic theory in the discovery of the ovaries (or female testicles) and argued that female and male semen contributed equally to both the form and the matter of the fetus and that reproduction was the result of the intermixing of the two sperms.[78]

Ibn Sīnā's theory of reproduction departed from the Aristotelian theory and attempted to modify it and reconcile it with the Galenic theory. Unlike Aristotle, Ibn Sīnā accepted the argument that women produce semen and that the female emission represented the basic female contribution to generation. Unlike Galen, however, Ibn Sīnā viewed female semen the same way that Aristotle did. He argued for the common origin of semen and menses; female semen, he said, was concocted menstrual blood. To be sure, menstrual blood provided nourishment for the fetus once it formed in the womb, but formation of the fetus was itself contingent on the mixing of male and female semen. Inception, according to Ibn Sīnā, was similar to the coagulation of cheese: the role of the female sperm in the formation of the fetus was similar to the role of milk in the formation of cheese, whereas the male sperm had the effect of the yeast that fermented the milk. Although "each of the two sperms is part of the essence of the fetus that is generated from them," the initiation of the thickening process is caused by the sperm of the man, which is the actor or agent, whereas the substance that is acted upon and thickens, ultimately to be transformed into a fetus, is the sperm of the woman.

Despite this obvious departure from Aristotle, Ibn Sīnā insists on pointing out the difference between his theory and Galen's. The point of contention is over Galen's argument that each of the male and female sperms is at once an agent and an object of fermentation (*'āqida wa qābila lil-in'iqād*). Significantly, however, Ibn Sīnā states that one could still accept Galen's view while assigning a stronger fermenting power to the sperm of the man and a larger receptivity to fermentation to the sperm of the woman. Ibn Sīnā, therefore, accepts the existence of male and female sperms and attributes to both of them active and receptive potentials, but he argues that the active agent that promotes fermenting is higher in the sperm of the man, whereas the passive potential of being fermented prevails in the sperm of woman.[79] According to this Aristotelian interpretation, the emission of the woman is not sperm. But to redeem Aristotle while subscribing to the Galenic view, Ibn Sīnā maintains that the sperm of the man is hot and thick, whereas the woman's sperm is of the same species as menstrual blood, which is why Aristotle calls it menses. In effect, Ibn Sīnā vindicates Aristotle by explaining him away but without denying the existence of female sperm. Despite his firm commitment to an Aristotelian natural philosophy, he holds that there can be no inception without the emission of the woman, that is, without orgasm (*idhā lam tanẓil lam yakun walad*).[80]

In contrast to Ibn Sīnā, Ibn Rushd attempted to redeem the Aristotelian theory in its totality and without modification. He insisted that even if women had testicles, and even if they produced semen, their semen was totally irrelevant to conception and played no role whatsoever in it. Ibn Rushd thus did not deny the existence of female semen, but he was categorical in rejecting the theory that female semen played any role in the impregnation of the woman or the formation of the embryo. The formative role in impregnation belonged exclusively to the sperm of the man. Neither the womb nor the female semen contributed to the process of procreation or to the creation of any of the organs of the fetus. Their only role was to preserve the semen of the man and prevent it from spoiling when exposed to air.[81]

Despite Ibn Sīnā's unwavering Aristotelianism, he allowed the findings of the practical discipline of medicine to take precedence over principles imposed on it from natural philosophy. Ibn

Rushd took the opposite approach, as we can see from the title of his book: *Al-Kulliyāt fī al-Ṭibb* (Universal [Principles] of Medicine). His intention in composing it was to reassert the superiority of outside universal principles over the knowledge that is internal to the discipline of medicine.

Back to Heaven: Astronomy

Even those astronomers in the eastern Islamic reform tradition who did not deny a hierarchy of knowledge effectively operated on the assumption that there were no insurmountable obstacles between various levels of knowledge, between perception and rationalization, or, in astronomy, between the mathematical and the natural philosophical. From Al-Bīrūnī's perspective, heavenly objects did not have an inherent nature, and even if they did, it was not relevant to the science of astronomy. Ibn al-Haytham, on the other hand, recognized that they did have an inherent nature and insisted on the need to reconcile mathematical astronomy and the natural science. But neither astronomer wanted to subjugate the science of astronomy ('ilm al-hay'a) to natural science (al-'ilm al-ṭabī'ī). Ibn al-Haytham did not offer solutions, but his and Al-Bīrūnī's critiques, each in its own way, meant that Ptolemaic assumptions could no longer be taken for granted. Theoretical astronomers no longer had a clear set of standards connecting mathematical astronomy to natural science—to use the language of Kuhn, they no longer had a "firm paradigm"—and this sense of a problem became one of the defining constants in the practice of all theoretical astronomers for a few centuries after the eleventh. The proposed solutions for this problem were diverse, however, not just in their mathematical technicalities but, more fundamentally, in what they attempted to achieve.

We have seen how Al-Qushjī and Al-Khafrī, like Al-Bīrūnī before them, disregarded natural science in their astronomical work. But what about astronomers who did not give up on the possibility of reconciliation between theory and reality? One of the main astronomers of the eastern reform tradition was Al-'Urḍī, a member of the Marāghā team.[82] I will focus on Al-'Urḍī partly because George Saliba, to whom we owe the most comprehensive comparative analysis of the planetary theories of the eastern reform school, has argued that the "outcome" of the astronomical reform tradition "can be summarized as returning the science of astronomy to its natural framework"—that is, to conformity with natural science—"and laying the mathematical foundation on which it must be built."[83] I, however, would like to suggest a different reading of the relationship between mathematical astronomy and natural science in Al-'Urḍī's work and, more generally, in the works of those astronomers who pursued Ibn al-Haytham's research program.

In the opening section of *Kitāb al-Hay'a*, Al-'Urḍī argues that the virtue of the science of astronomy derives from the nobility of its subject matter, the heavenly bodies, and from the certainty of its "geometric, mathematical" proofs (27–28). Significantly, Al-'Urḍī does not mention natural science in this introduction. Later, however, he includes several sections in which he discusses postulates of Aristotelian natural science that relate to astronomy, including the principle that heavenly bodies are neither light nor heavy, from which follows the circularity of the motion of these bodies, the impossibility of vacuum, and so on (29–34). Following these clear but brief introductions, Al-'Urḍī proceeds to prove some of these axiomatics. The proofs he provides are mathematical and observational, and none follows the standard proofs that would be given in natural science.[84] His choice of proofs is in no way accidental. In one instance, Al-'Urḍī refers to people who, on account of their lack of knowledge of the mathematical sciences (*al-'ulūm al-ta'ālīmiyya*), base their views on principles external to the discipline of astronomy and assert that all the orbs have to move in one direction. In response, Al-'Urḍī maintains that "observation and mathematical reasoning require (*al-i'tibār bil-raṣd wal-naẓar al-ta'līmī iqtadā*) that the motion of the planets is compound" (48–49). Here, Al-'Urḍī resorts to mathematical reasoning and observation to undermine a principle of Aristotelian natural philosophy.

Once Al-'Urḍī commences his analysis of planetary models, he no longer refers to natural science (ṭabī'ī) and refers only to principles (uṣūl) internal to the discipline—specifically to the principle that spheres must have uniform circular motions around their own centers or, conversely, that a sphere cannot have a uniform circular motion around a point that does not coincide with its center. Invariably, this is what Al-'Urḍī means when he refers to principles, and all his references to uṣūl are minor variations on this theme.[85] To be sure, uniform circular motion is one principle of natural philosophy, but as we have seen from the accurate understanding of Aristotelian natural philosophy by the philosophers of the western Islamic tradition, this is only one of many other principles ignored by Al-'Urḍī and other astronomers of the eastern Islamic tradition. In Al-'Urḍī's work this principle is taken out of its natural philosophical context and given an exclusively mathematical flavor. Additionally, despite Al-'Urḍī's contention that there needs to be some sort of correspondence between the mathematical models and reality, in much of his work he applies the principle of uniform circular motion to imagined circles (*dawā'ir mutawahhama*).[86] In other words, Al-'Urḍī's use of uṣūl does not necessarily entail the realism of the proposed models for planetary motions.

Similarly, most of Al-'Urḍī's discussions of problems in Ptolemaic astronomy are about discrepancies between the mathematical models and the uṣūl of astronomy or the uṣūl that Ptolemy established for this science. In contrast to Ibn al-Haytham's initial formulations, Al-'Urḍī's discussions hardly refer at all to a contradiction between mathematical astronomy and natural science (al-'ilm al-ṭabī'ī). In one exception that seems to confirm the significance of his careful choice of words, Al-'Urḍī refers to the divergence of Ptolemy's lunar model from what Ptolemy "thought is the nature of the heavenly bodies" (*al-amr alladhī yarā annahu ṭabī'at al-ajrām al-samāwiyya;* 115). Significantly, Al-'Urḍī does not assert here an absolute nature but only an estimation of what this nature might be.[87] In the same discussion, Al-'Urḍī explains the reason for Ptolemy's failure: "I think that the sciences—in fact, even the crafts—are seldom found complete in their beginnings, and they gradually become more perfect." That is, the principles of science evolve as a science itself evolves (115, 228). In order not to leave any ambiguity, Al-'Urḍī explains the objectives of his alternative project: "to search for the truth [of planetary motions] and to imagine for them configurations (hay'as) that conform to the rules set as principles for this science [ilm al-hya'a]." No science, Al-'Urḍī adds, can dispense with principles that either have their proofs outside the science or are self-evident facts that are made into principles for that science. The self-evident principles that the "mathematical (*ta'ālīmī*) [astronomer] receives as givens on which to base his work are that the heavenly motions are circular and uniform and that the observations and the average values derived from the rotations are what they are" (115–16).

Having established the principles, Al-'Urḍī proposes an alternative to the Ptolemaic lunar model.[88] In Al-'Urḍī's model, the center of the epicycle is made to move in a direction opposite to the one assigned to it by Ptolemy, and the value of this motion is three times the Ptolemaic value. Al-'Urḍī justifies this remarkable reversal of direction and the change in value as follows: "Our objective is that the outcome (*ḥāṣil*) from our model and his [Ptolemy's] are the same and correspond to what was apparent to him in reality (*ma ẓahara lahu bil-ḥaqīqa*), and to abide by matters that follow the right method—that is, conform to the uṣūl" (116). The "reality" that Al-'Urḍī is concerned with here is the observational reality, the visible location of the planet, and not the reality defined in natural philosophy. The principle of all principles that Al-'Urḍī never tires of repeating and on the basis of which all models should be constructed is the principle of a uniform circular motion around a sphere's center. Nowhere in his alternative models does Al-'Urḍī deploy the natural philosophical principles outlined in the introductory section of his book.

On several occasions Al-'Urḍī refers to configurations or planetary models that are "in conformity with the principles thought natural to the heavens" or that are "more appropriate representations of the nature of the heavens (*'alā wajh alyaq wa ashbah bil-ṭabī'a al-samāwiyya*)" (188, 219). These natural principles are, however, a far cry from the principles that have their certain demonstrative proofs in natural science. Al-'Urḍī, therefore, does subscribe to Aristotelian natural philosophy, but what he extracts from it is far removed from what natural philosophers would. In his work, mathematical proofs completely marginalize natural philosophical arguments; the one principle retained from natural philosophy, the principle of uniform circular motions, is removed from its original natural philosophical context and given a mathematical flavor; and realism, or correspondence to reality, is no longer dogmatic but mathematical. What I mean by this last point is that while Al-'Urḍī recognizes the need for models to correspond as much as possible to outside reality, there is no evidence in his models that he believed that they had a definite identity with reality. In fact, in the lunar model mentioned above, to give just one example, Al-ʿUrḍī does not hesitate to reverse the direction of the motion and to triple its value. It is hard to see how Al-ʿUrḍī could possibly have conceived of this reversal as one that corresponds to reality. As he tells us, however, his model does allow for an accurate prediction of the final (*al-ḥāṣil*) planetary positions as provided by Ptolemaic observations while employing spheres that rotate uniformly around their own centers. We would likewise be hard pressed to provide natural philosophical explanations or to identify real physical counterparts for mathematical tools such as Al-Ṭūsī's couple and Al-ʿUrḍī's lemma, both of which are called uṣūl by Al-Shīrāzī.

Although many astronomers in this tradition tried to make a case for one or another of their favorite models, no model ultimately triumphed. Had the reality of the planetary models been an absolute requirement, there would be only one correct model for each planet, and its correctness would be established by absolute demonstrative rules derived in natural philosophy.[89] The "nature" that mathematical astronomers tried to conform to was no longer the nature defined in Aristotelian natural science. But this does not mean that the new astronomy had no guiding principles or that by abandoning strict adherence to Aristotelian physics, astronomy was moving into a vacuum (as it were).

Even when the astronomers continued to propose solid body representations for planetary movements, presumably to match the way the orbs would appear in reality, the validity of their models was grounded not in natural philosophical reality but in conformity to the mathematical principle of uniform motions, a principle that was gradually removed from its original natural philosophical understanding. What started as an attempt to restore harmony between mathematics and physics resulted in mathematical astronomy proceeding in almost total disregard of physics. A new hierarchy of principles was thus established in which mathematics provided the primary source for the coherence of the discipline of astronomy. In the new astronomy, mathematical structures imposed their logic on physical reality, and mathematical principles were tools not just for studying nature but also for conceptualizing it.

Already toward the end of the twelfth century, Al-Bitrūjī (of the western Islamic astronomical tradition) had captured this fundamental characteristic of the eastern astronomical tradition. In his *Kitāb al-Hya'a* he says: "The mathematical astronomers (*aṣḥāb al-ta'ālīm*) reflect on what they perceive with the senses about the conditions [of the celestial bodies] and turn it [what they perceive] into a principle (aṣl). They depend [in their definition of principles] on what perception entails in regard to the motions of the planets, and they disregard what the intellect entails and what their [the planets'] natures require. This they do even while they concede that most of the motions of the planets that are perceived by the senses are different from the reality [of their motions]."[90] Al-Bitrūjī, who wanted to bring natural philosophy back into astronomy, rightly identified the key to the eastern astronomical reform tradition: the formulation of principles from within the discipline, or, put differently, the assertion of the autonomy of the science of astronomy vis-à-vis natural philosophy.

The novelty and the epistemological coherence of the new astronomy derived above all from the conceptual separation between science and philosophy. No longer was the organic unity of knowledge assumed. The sciences did not stand in complete isolation from each other, but they derived their guiding epistemological principles primarily from within their own shared practices and assumptions. The old Aristotelian cosmos gradually fell apart as scientists began to search for alternative principles of coherence in their respective disciplines. Admittedly, this search did not produce a unified and unifying alternative scientific cosmology. But the internal coherence of the various disciplines proved to be a much more constructive force in their development than the metaphysical unity that they lost. Now that the configuration of the universe and the causes of the motions of heavenly bodies were no longer intertwined, astronomers were free to search for mathematical representations without having to worry about first causes. Of course, many philosopher-scientists continued to try to track down first causes, but the overall practice of science was no longer bound to their pursuit.[91]

Classifying Knowledge and the Place of Reason

The debates examined in this chapter are about ways of knowing and the relationship between various systems of knowledge. Theories of knowledge were often elaborated in classifications of science.[92] The classifications genre proliferated in Muslim culture, and whether directly or indirectly, all classifications dealt with the position of various disciplines of learning relative to each other and the relationship between religious knowledge and scientific knowledge and knowledge more broadly. Some contemporary historians have had difficulty in identifying patterns in the rapidly evolving genre of classification and have as a result attributed their confusion to the genre itself. But since the genre's subject was knowledge, and since knowledge was rapidly evolving in Muslim societies, the genre needed to redefine its own subject constantly. With every new discovery in science, with the new understanding of existing sciences, and with the invention of new sciences, the matrix of relationships between the various sciences changed, making the genre of classification incomplete by definition.

To be sure, the organizing principles of classification differed. In addition to the standard Aristotelian classification of the sciences into productive, practical, and theoretical, there were classifications based on the mode of acquiring knowledge or on the subject matter.[93] Other classifications were based on the kind of knowledge obtained (e.g., theoretical versus practical), the utility of the sciences, and so on. Irrespective of method, all classifications shared the objective of discerning order in the seemingly orderless pursuit of knowledge. They did not just try to account for the varieties of human knowledge but also provided a blueprint for achievable knowledge, thereby suggesting what had been and remained to be achieved, along with ways of achieving it.

I will not attempt a comprehensive overview of the genre of classification or even of an individual classification scheme. Instead, after sampling key features of various theories of knowledge, I will use the classifications provided in the work of the celebrated historian Ibn Khaldūn (d. 1406) as a window into the larger debates in the context of which the relationship between science and philosophy was negotiated. Ibn Khaldūn's introduction to history (*Al-Muqaddima*), which was an attempt to turn the craft of writing history into a science, defies easy attempts to simplify it. About a third of the text contains epistemological assessments of the various sciences. Ibn Khaldūn's classifications are quintessential epistemological studies that address the premises of the sciences, their assumptions, valid subject matter, appropriate questions, and results, as well as modes and conditions of scientific knowledge, the truth value of the knowledge that derives from different sciences, and the limits of knowledge within each of the sciences. In some respects, Ibn Khaldūn's text can also be read as a history of science and of the relationships between the different sciences. The complexity of the text, I think, is also significant in its own right, because it reflects the historical state of the relationship between the sciences and the multiple conceptions and epistemologies underlying them. So I read Ibn Khaldūn both as a repository and as an outcome of the historical debates on science and epistemology.

Although the sciences were initially referred to as the ancient sciences, with the passage of time the term "rational sciences" became more common. Ibn Khaldūn also used the expression "the sciences shared among all nations." Irrespective of name, however, the place of reason is central to the conception of all theories of knowledge and all classifications of science. The word for "reason" in Arabic is *'aql,* which, depending on context, translates as "mind" or, in the philosophical usage, "intellect." The different translations reflect two broad generic distinctions in the understanding of reason. I will call the types of reason metaphysical reason and procedural reason.

The earliest theoretical discussions of 'aql were in the context of the philosophical tradition. Ibn Sīnā provides a typical understanding of the term, distinguishing between the everyday usage of the term and the more technical usage by philosophers. In philosophical understanding, the term has eight possible meanings, the highest of which, according to Ibn Sīnā, is "active intellect" (*al-'aql a-fa'āl*), which is a quiddity (*māhiyya,* "essence") completely abstracted from matter (*mujarrada 'an al-mādda aṣlan*).[94] This active intellect of the philosophers does not derive its reasoning ability, its intellections, from material beings; instead, it intellegizes (*ya'qul*) the totality of existence. The "human intellect," in comparison, acquires its building blocks (*ma'qūlāt,* "intelligibles") by abstracting them from material experience, and the increase in abstraction gradually elevates the human intellect toward the perfect active intellect.

Generally speaking, the philosophical classifications of science were made either on the basis of subject matter or methods, or on the basis of the kind of knowledge attainable in the sciences. The hierarchy of knowledge produced by the philosophical conceptions of reason took status partly from the nobility of the subject of study but mainly from the kind of intellection employed in studying a subject. Thus, theoretical knowledge was higher than practical knowledge, and the highest form of theoretical knowledge was *kullī*, knowledge, which dealt with universals whose subject was being qua being (*al-jawhar bimā huwa jawhar, al-wujūd bimā huwa wujūd*), not perceptible individualized beings (*jawhar maḥsūs*). According to Ibn Sīnā, this First Philosophy, as it were, included divine philosophy (*al-ʿilm al-ilāhī* or *ilāhiyyāt*), as well as the science that deals with the principles or premises (*mabādi'*) of partial sciences like mathematics and logic. Not only is theoretical knowledge of universals higher than the knowledge that obtains in the partial sciences, but the science of the universals (*al-ʿilm al-kullī*) is the only science that provides demonstrative proofs for the principles of the partial sciences. The ultimate science, therefore, is the universal science, the science of the first causes and the absolute principles of all sciences; all partial sciences are therefore ontologically dependent on the universal science.[95] Ibn Sīnā argues that the universal principles (*mabādi' kulliyya*) of physics are conventional insofar as it is a partial science; however, these principles are demonstrated in the universal science, in metaphysics and divine philosophy, and practitioners of the natural science (al-ʿilm al-ṭabīʿī) simply receive these principles and do not prove them inside their own science.[96]

What we have in this traditional philosophical schema is a notion of a metaphysical intellect that has an existence separate from human intellection. And parallel to this metaphysical intellect, we also have kinds of universal theoretical knowledge, kulliyāt, that bind and override the forms of human reasoning that are exercised in the partial sciences.

A variety of views in contrast to the philosophical notions of ʿaql were expressed across the spectrum of Islamic thought, but they tended to converge on a procedural approach to reason. Members of the two main theological schools, the Muʿtazilīs and the Ashʿarīs, considered reason not an essence but a tool, or, alternatively, the accumulation of individual knowledge that enables a person to discriminate between things and to be able to acquire additional knowledge through reflection and inference.[97] The religious scholar Al-Ghazālī, for example, maintained that ʿaql was a measure and a balance used to distinguish between things and to determine the extent of the correctness of things imagined or perceived by the senses.[98] Along the same lines, Ibn Taymiyya (d. 1328) maintained that "the fact that evidence (*dalīl*) is rational or [obtained] through transmission is not in itself a trait that requires praise or criticism or that [establishes whether] it is true or false. Rather, it shows the way by which something is known, which is either through transmission or reasoning, although by necessity transmission is always in need of reason. . . . As for it (dalīl) being lawful (*sahrʿī*), this [trait] is not contrasted with it being rational but with being innovational, since innovation is the opposite of lawfulness."[99] To Ibn Khaldūn, a final example in this brief list, reason provides tools and methods of thinking relevant to specific sciences or fields of knowledge; and reason builds itself up by using the combination of tools available to it at any particular moment. So, instead of assuming an absolute intellect with a real existence, philosophers in this tradition shifted their focus to trying to understand how humans build their intellectual faculty or their reason. ʿAql was no longer an essence, then, but a faculty, and the emphasis shifted from identifying reason to identifying what is rational.

In Ibn Khaldūn's treatment, this practical understanding of reason limited its scope and left certain subjects outside its jurisdiction. He says: "Perhaps there is a kind of perception other than what we [humans] have, because our perceptions are created and accidental, and the creation of God is larger than the creation of humans. . . . This [view] does not belittle reason and the perception of reason in any way, because reason remains a truthful balance, and its judgments are certain and free of falsehood. However, you cannot hope to weigh with it matters such as the oneness [of God] and the hereafter, nor can you determine through it the truth regarding prophethood, divine traits, or anything else that lies outside its domain, for that would be to desire the impossible."[100]

Practical reason, therefore, does not stand outside the field in which it is generated. Ibn Taymiyya sums up this view eloquently:

> That which is an absolute universal in the minds (*adhhān*) of people can be found only as a particular embodied, specific, and distinct thing in reality. It is referred to as a universal only because it is such in the mind (*fī al-dhihn kullī*). As for external [existence], there is nothing in it that is universal. This principle applies to all the sciences. . . . We know through perception and by rational necessity that the outside world has nothing that is not individualized and of a particular designation (*muʿayyan mukhtaṣṣ*), and there is no sharing in it at all. However, all the generalized universal meanings in the mind (*dhihn*) are similar to absolute and general utterances made by the tongue, or to the script that signifies these utterances. The script corresponds to the utterance, and

> the utterance corresponds to the meaning. Each of the three relates to the individual things that have an external existence [i.e., external to the mind] and encompasses them and applies to them, but this does not mean that something that encompasses, or is inside, or shares in this and that [individual entity] has an external existence.[101]

To Ibn Taymiyya, universals were mental constructs that had no real existence. It followed that such universals could not provide the principles of specific sciences. Ibn Taymiyya's disciple Ibn Qayyim al-Jawziyya (d. 1350) says of philosophers that "they abstracted universals (*umūr kulliyya*) that have no external existence, applied to them the rules of real beings (mawjūdāt), and then made them the measures and principles of existing things."[102]

Many of the views discussed so far are those of religious scholars, but their critiques of metaphysical reason were conducted on epistemological grounds and were shared by many scientists who argued the autonomy of their disciplines and sought to devise the principles of their sciences from within their own internal practices. The continuing evolution of the classification of science was itself a reflection of the ongoing rethinking of epistemology in light of the real development of knowledge.

Like Ibn Taymiyya, Ibn Khaldūn interrogates in his classification the use of pure inductive reason in natural philosophy, for in this realm, reason has its limits and is in need of other kinds of proof. Inductive reason produces generalized universal mental judgments, whereas beings that are external to the mind are individualized matter (*mutashakhkhiṣa bi mawāddihā*), and this matter might have in it that which prevents the "identity between the universal mental and the individualized external." Significantly, Ibn Khaldūn does not deny the validity of abstract mental judgments in certain realms.[103] Divine philosophy and metaphysics (ilāhiyyāt) are not included in those realms, for their "essences are unknown to start with, and we cannot reach them or prove them, because it is only possible to abstract intelligibles from external individualized beings for things that are perceptible to us." Spiritual essences can be conceptualized only after abstracting them from material existence; therefore, no certain demonstrative proof could apply to them either, for, as the leading philosophers have maintained, "a thing that does not have matter [material existence perceptible by the senses] cannot be proved by demonstration (*mā lā mādata lahu lā yumkin al-burhān ʿalayh*)."[104]

For Ibn Khaldūn, therefore, the universal principles that derive from an abstracted metaphysical intellect do not apply to natural knowledge, which deals with things that are perceptible by the senses. If, however, there is a direct correspondence between reasoning and that which is perceived by the senses, then it is possible to accept the judgment of this procedural reason in natural science. Reason, therefore, is operative only when applied to individualized material beings; it does not apply to anything beyond material existence. Again, what is meant by reason is a set of principles that, when systematically applied, provide certain knowledge in a particular science.

One of the fundamental consequences of Ibn Khaldūn's notion of practical or procedural reason is that the validity of a form of knowledge is predicated no longer on a moral judgment but on the correspondence between its subject matter and the mode of reasoning exercised in dealing with that subject matter.[105] Ibn Khaldūn's classification is grounded in a historical view of science as a cultural artifact (*min maẓāhir al-ʿumrān*), as a product of the cumulative practical and theoretical activities of humankind. As such, science is no longer seen as the gradual perception of natures that have prior existence to its historical practice. Heavens and metaphysics, in the disciplinary philosophical sense of the words, are not determining factors in either the existence or the production of scientific knowledge. Leaving no room for ambiguity, Ibn Khaldūn maintains that "experiential skill engenders intellection (or 'reason' or 'intellect'; *al-ḥinka fī al-tajriba tufīdu 'aqlan*); technical faculties engender intellection; and civilization as a whole (*al-ḥadāra al-kāmila*) engenders intellection. Because these things are the cumulative outcome of crafts (ṣanā'i') . . . and all of these are rules (*qawānīn*) that cohere in sciences (*tantaẓimu 'ulūman*), 'aql increases as a result (*yaḥṣul minhā ziyādat al-'aql*)."[106] Manual labor, according to Ibn Khaldūn, involves a kind of theoretical reasoning because it presupposes the ability to conceptualize external reality in a particular natural or conventional order before executing a task in this order; in fact, "all crafts are faculties related to practical mental matters (*al-ṣīnā'a hiya malaka fī amr fikrī 'amalī*)."[107]

Thus, to Ibn Khaldūn, science and knowledge are incremental. In one of many outstanding insights, he uses this understanding of science to distinguish between thinkers in the eastern and western parts of the Muslim world. The virtue of the easterners over the westerners, he notes, is on account of their "receptiveness to knowledge that occurs to the soul as a by-product of civilization through the instrument of incremental reason (*al-'aql al-mazīd*)."[108]

To sum up, reason builds itself in history and has no metaphysical existence, according to Ibn Khaldūn. The abstract reasoning posited for divine philosophy and metaphysics is not possible because the subject matter of these sciences has no tangible material existence that can be perceived by the intellect.[109] The universal principles of these sciences are not suitable for their own subject matter, let alone for other sciences. In fact, Ibn Khaldūn carried this epistemological argument to its logical end when he rejected the assertion by Muslim philosophers that prophets are a logical necessity for human societies. Religion is not a logical necessity, he believed, and social and political life does not need religions or prophets, as countless forms of social order not predicated on religion attest. Ibn Khaldūn was not denying the benefits of religion, but he maintained that its existence was simply a gift from God and not a consequence of the universal principles of the metaphysical sciences.[110]

By determining those parts of the world that are subject to rational reflection, Ibn Khaldūn liberated scientists from the burden of attempting to explore what is not knowable. Of course, he was simply articulating what scientists since Al-Bīrūnī had been doing in their respective fields. In the mainstream of Islamic scientific culture, the natural philosophical and metaphysical principles that unified the sciences of antiquity were eroded. The various sciences operated, not on the assumption of a unifying universal reason, but on the assumption that the criteria of rationality were not independent from the intellectual and historical contexts of the disciplines in which rationality was deployed.

Chapter 3

Science and Religion

Some important debates that pertained to the nature of scientific knowledge, the intellectual authority of science, and the social and institutional authority of scientists were primarily religious, not merely joined by religious scholars or driven by a sense of religious unease about the sciences. It is of course a truism to say that the relationship between science and religion in Muslim societies is multifaceted. Nothing else could have been the case, given the temporal and geographical spans of the scientific enterprise in these societies and the diversity of the intellectual traditions that informed views about the science-religion relationship. Although a major trend in medieval Islamic culture was to argue the autonomy of discrete forms of knowledge, science as an intellectual construct could not have constituted a neutral system of axioms and laws. Asserting distinction was itself part of a meaningful epistemological project, but such assertions were not necessarily objective descriptions of the reality of scientific or religious thought as socially and culturally embedded phenomena. Even attempts to delineate orderly distinctions between science and faith have been subverted by the constant transformation of each of these forms of knowledge. Religious and scientific knowledge are constantly reconstituted within different social and cultural contexts, and yet anyone making a nondogmatic examination of either over the course of time would recognize that the two are always intertwined in actual history. Attempts to establish the autonomy of scientific or religious knowledge were deliberate undertakings that informed historical developments in these two areas and the way they interacted with each other; however, the very attempts to assert a distinction were themselves reflections of particular forms of historical and epistemological communication between science and religion.[1]

Although the traditional Orientalist assertion of a fundamental opposition between science and religion is, to put it mildly, no longer tenable, the notion of inherent harmony and unity is as problematic as the opposite notion of inherent conflict. In the previous chapter I argued that there were multiple and at times conflicting epistemological assumptions in the practice of science. Religious discourse on science was equally diverse, running the whole gamut from assertions of opposition and conflict to assertions of harmony and unity. Both religious opposition to and endorsement of science took different forms, depending in part on whether the basis was doctrinal reasoning, notions of utility or uselessness, or epistemological criticism or sanction.

Science and religion, the rational and the sacred, intersected in numerous cultural spaces. At formal levels, science and religion interacted socially and institutionally. Scientific and religious practices were combined in the persons of countless numbers of scholars who specialized in various rational and religious sciences, and the lines separating religious and scientific scholars were often blurred. Numerous frameworks instituted to nurture and support the practice of science paralleled religious institutions in structure. Like communities of religious scholars, scientists also asserted collective professional identities and tried, in a variety of ways, to secure social sanction for their communities. By focusing attention on the specific institutional and professional interests of individuals or groups, these forms of intersection could easily account for harmony or for conflict between religion and science. Among other things, the notion of conflict was often rhetorically deployed to serve particular agendas and to support the claims of competing professional groups or individuals. While these motivations are important, they are by definition specific to time and place; moreover, they technically fall in the realm of the sociology of knowledge and cannot be generalized to account for epistemological developments over long periods of time.

Scientific and religious disciplines did intersect, as we have noted with the sciences of mīqāt (timekeeping) and farā'iḍ (inheritance algebra). All of the forms of interaction were circumstantial, however, and do not account for the conceptual interaction between science and religion. Here I will trace, in the realm of religious thought, trends that correlate to the epistemological developments in the realm of science. To be sure, my starting assumption is that a scientific culture that had a vibrant life for at least seven or eight centuries could not have flourished in isolation from other cultural forces, especially religion. In other words, Islamic science developed in the context of Islamic culture and not despite this culture, as many historians have asserted; specific developments in religious thought corresponded to and reinforced the conceptual developments in scientific thought.

Where do we look? At a basic level, it is easy to see how certain religious views could provide indirect positive influences on science. We often encounter in scientific literature, especially in the opening sections of scientific treatises, references to the Qur'ānic call on Muslims to reflect on the outside world and the world within, where they would find evidence of God's flawless creation and bounty. Knowledge, in this view, is encouraged because it deepens piety, but also because it is considered a virtue in its own right.[2] Certain kinds of knowledge are also constitute collective religious obligations. In contrast to religious obligations required of individual Muslims (*fard 'ayn*), such as praying and fasting, a second category of obligations (*fard kifāya*) are obligations of which individuals are absolved as long as there are enough people in the community to fulfill them—they include all crafts and all kinds of knowledge on which the survival and well-being of a community depend. For example, the community must have enough physicians to attend to the health of its members, enough grammarians to understand the Qur'ān, and so forth. These and other organizational aspects of religion, including creed, ritual, and institutional context, helped shape the practical attitude of religion toward science. Still, we also find arguments to the effect that all the knowledge needed by Muslims is in the Qur'ān and that even if the knowledge found in the rational sciences is correct, it is useless. These views, however, tell us very little about the sciences, their social status, or their epistemological assumptions. They are simply a by-product of the complex modes of interaction between two living cultures: science and religion.

Beyond straightforward acceptance or rejection of science, we can also recognize the possible role of religious and theological debates, both within the Islamic community and between Muslims and non-Muslims, in motivating Muslims to seek intellectual ammunition in the arsenal of ideas available to them. The most obvious source was the legacy of Greek philosophy that had already shaped Christian theology, and Muslim theologians drew on Greek sources as they developed their own theological arguments.[3] Yet this possible theological context for the interest in the Greek philosophical legacy did not predetermine, nor does it explain, the subsequent religious attitude toward science.

In some instances, religious doctrine had a direct effect on the shaping of scientific practice. For example, many scientists, philosophers, and religious scholars criticized astrology, an integral component of the Greek scientific legacy, on epistemological grounds but also because it contradicted the religious doctrines of Islam. As George Saliba argues, the emergence of the new science of hay'a (theoretical astronomy) was a specific response to the religious attacks on Greek astrology; in the face of these attacks, Islamic astronomers consciously redefined and reoriented their discipline, and in so doing they created a new discipline with no equivalent in the Greek tradition.[4] As a result, although astrology continued to be practiced both at the popular level and among the political elite, "remarkably little astrology" is preserved from the classical period of Islam—while thousands of scientific treatises survive.[5]

Another gauge of the relationship between science and religion is the way scientists perceived of their practices in relation to religious knowledge. Al-Bīrūnī, an eminent representative of the scientific tradition, may have also recapitulated in his science trends in religious thinking about science. In his anthropological history of India, Al-Bīrūnī starts a chapter entitled "On the Configuration of the Heavens and the Earth According to [Indian] Astrologers" with a long comparison between the cultural imperatives of Muslim and Indian sciences. The views of Indian astrologers, Al-Bīrūnī maintains, "have developed in a way that is different from the ways of our [Muslim] fellows; this is because, unlike the scriptures revealed before it, the Qur'ān does not articulate on this subject [of astronomy], or on any other [field of] necessary [knowledge], any assertion that would require erratic interpretations in order to harmonize it with that which is known by necessity." The Qur'ān, adds Al-Bīrūnī, does not speak on matters open to endless differences of opinion, such as history. To be sure, Islam has suffered from people who claimed to be Muslims while retaining many of the teachings of earlier religions and claiming that these teachings were part of the doctrines of Islam. Such, for example, were the Manichaeans, whose religious doctrine, together with their erroneous views about the heavens, were wrongly attributed to Islam. Attributions of such scientific views to the Qur'ān were, according to Al-Bīrūnī, false claims of un-Islamic origin. The opposite took place with the Indians, all of whose religious and transmitted books speak "of the configuration of the universe in a way that contradicts the truth known to their own astrologers." But, driven by the need to uphold their religious traditions, Indian astrologers pretended to believe in the astrological doctrines of these books despite being aware of their falsity. With the passage of time, accurate astronomical doctrines were mixed with those advanced in the religious books, which led to the confusion encountered in Indian astronomy.[6]

Although not all Indian religious views contradict the findings of the astronomical profession, says Al-Bīrūnī, the conflation of religious and astronomical knowledge undermines Indian astronomy and accounts for its errors and weaknesses. He contrasts this conflation of scripture and science with the Islamic astronomical tradition, which, in his view, suffers from no such shortcomings: the Qur'ān, and therefore religion more broadly, does not interfere in the business of science, nor does science infringe on the realm of religion. As we shall soon see, Al-Bīrūnī was not off the mark in his assessment of the Qur'ānic attitude toward science.

Perhaps more relevant for our purposes are the systematic discussions and epistemological assessments of science by religious scholars and by Muslim intellectuals. Numerous important religious thinkers, such as Ibn Ḥazm (d. 1064), Abū Ḥāmid Al-Ghazālī, and Ibn Taymiyya, dedicated significant portions of their writings to the discussion of various sciences, scientific knowledge, and the relation of scientific knowledge to religious knowledge. Religious scholars wrote many of the classifications of science, which reflect a diversity of assessments. Some classifications amount to simple lists of the rational sciences and the religious/traditional sciences in two separate categories without any comments on their relationship or relative value. Other classifications underscore the utility of the rational sciences to the religious sciences as a way of legitimating them. Still others compare parallel structural features that relate some religious sciences to their rational counterparts, whether by the mode of acquiring knowledge or by the nature of the science, practical verses theoretical. In addition, many classifications provide critical assessments of various scientific disciplines on ethical as well as epistemological grounds.

We can go on listing many more areas of contact between science and religion, but rather than marshaling the testimonies of religious scholars or works in support of science (or, for that matter, in opposition to it), the question for us is whether we can trace general trends in individual attitudes. To be able to generalize, we need to focus on the salient treatments of science and scientific knowledge in religious writings, not just the way communities of scientific knowledge conceived of their professions and research within the larger context of religion but the way scientific knowledge was situated within religious discourse. In what follows, I will address this question by examining two primary genres of religious scholarship: *tafsīr*, or Qur'ānic exegesis, and *kalām*, or speculative theology. I will also briefly examine the religious stand on causality, a subject of interest to many medieval Muslim scholars and to contemporary historians of Islamic science as well.

Tafsīr

The Qur'ānic attitude toward science, in fact, the very relationship between the two, is not readily identifiable. Almost all sources agree that the Qur'ān condones, even encourages, the acquisition of science and scientific knowledge and urges people to reflect on the natural phenomena as signs of God's creation. Beyond this simple contention, however, there is great disagreement over the religious position of science. Qur'ānic paradigms of science are articulated mainly in the genre of Qur'ānic exegesis (*tafsīr*, plural *tafāsīr*). As much as exegetical works insist on grounding themselves in the immutable text of the Qur'ān, they are also repositories of larger cultural debates and reflect the prevailing views at their time and place of writing. Of course, the genre of tafsīr is vast, and many exegetical works have nothing to say about science. So rather than tracing the evolution of the exegetical views on science, I will focus on some representative works in this tradition and try to identify their general thrust.

Traditional Qur'ānic exegetical works contain plenty of material of possible scientific import. Despite current interest in the Qur'ān and science, this aspect of exegesis has not received much scholarly attention. One possible reason for the neglect is that collectively, the traditional materials do not add up to what might be legitimately called a scientific interpretation of the Qur'ān. Traditional exegetes did not present themselves as engaging in such an interpretive exercise. Only a minority of traditional scholars, notably Abū Ḥāmid al-Ghazālī and Jalāl al-

Dīn al-Suyūṭī (d. 1505), maintained that the Qur'ān is a comprehensive source of knowledge, including scientific knowledge.[7] To support their contentions, Al-Ghazālī and Al-Suyūṭī cite such verses in the Qur'ān as "for We have revealed to you the Book as an exposition of everything" (Qur'ān 16:89). The same verse, let me note, starts with "Remind them of the Day when We shall call from every people a witness against them, and make you a witness over them" and, after describing the Qur'ān as an exposition of everything, goes on to say, "and as guidance and grace and happy tidings for those who submit." The likely reference in this verse to the exposition of knowledge is connected to knowledge of what will happen in the hereafter and to the fate of believers therein. Despite their claims, neither Al-Ghazālī nor Al-Suyūṭī proceeds to correlate the Qur'ānic text to science in a systematic, interpretive exercise. Nor are there instances in which these two or other exegetes claim authority in scientific subjects on account of their knowledge of the Qur'ān. Perhaps the most relevant reason for the absence of an articulation of a Qur'ānic paradigm of science in premodern times was that such an articulation was not needed: the counterclaims of a hegemonic culture of science and the ideological outlook that accompanied the rise of modern science were both absent.[8]

Scientific subjects do come up in many medieval Qur'ānic exegetical works, but their treatment is radically different in these sources than in today's counterparts. I will address the contemporary approach in the next chapter, but now I will explore the paradigmatic treatment of this subject in classical Qur'ānic exegeses. The scientific discourse in the classical Qur'ānic commentaries is invariably mixed with other kinds of discourse that have no connection to science. Qur'ān commentators had a distinct conception of what constitutes the main thematic emphasis of the Qur'ān, and they often, though not always, presented their detailed discussions of various subjects within this framework. Thus, for example, in a commentary on one verse (Qur'ān 7:54), Fakhr al-Dīn al-Rāzī (d. 1210) spells out the four themes around which the various discussions of the Qur'ān revolve (*madār amr al-Qur'ān*).[9] The verse in question, which relates to the natural order, reads: "Surely your Lord is God who created the heavens and the earth in six days, then assumed the throne. He covers up the day with night, which comes chasing it fast; and the sun and the moon and the stars are subjugated by His command. It is His to create and command. Blessed be God, the Lord of all the worlds."[10] Before embarking on a lengthy discussion of this verse, Al-Rāzī lists four overriding Qur'ānic themes: the oneness of God, prophethood, resurrection, and the omnipotence of God or the related question of predestination.[11] All other themes, including any others in this verse, underscore one of these four essential motifs.

Al-Rāzī explains the manner in which this seemingly unrelated verse does indeed relate to the oneness and omnipotence of God, and lists several interpretations of the verse that confirm this correlation. According to one interpretation, the heavens and the earth were created with a particular size, although their natures do not preclude the possibility of their having a larger or smaller size. This shows, says Al-Rāzī, that a willing maker chose to give them this specific size and no other, thus proving the existence of a free and willing creator (13–14:96–97). Alternatively, the creation of the heavens and the earth at a specific time, when they could have been created at an earlier or later time, is an act of choice by God and not due to the inherent nature of either creation. The same argument applies to the configurations and positions of the various parts of the universe relative to one another, and so on (97–98). After a lengthy digression to disprove the attribution of place and direction to God, Al-

Rāzī returns to the first theme, albeit from a different perspective (98 ff.). He enumerates the benefits that result from the succession of day and night, again as proof that God created the world in a specific fashion in order to maximize its benefits for humans (117). Al-Rāzī then undertakes a linguistic exploration, typical of Qur'ānic commentaries of all kinds, of the meaning of the word *musakhkharāt* (subjugated). The sun, he reports, has two motions: the cyclical rotation of one is completed in a year, that of the other in a day. Night and day, however, are due not to the motion of the sun but to the motion of the great orb that, according to Al-Rāzī, is also the throne (117–18). An angel is assigned to move each heavenly body or planet when it rises and sets (118–19), and God has endowed the throne, or the great outer orb, with the power to influence all the other orbs, enabling it to move them by compulsion from east to west, in the opposite direction of their slow west-to-east motion (119–20). This, says Al-Rāzī, is the meaning of subjugation: that orbs and planets are organized by God in a particular order, *for no inherent reason of their own*, so that they produce the optimal benefits for humans (120; my emphasis).

Al-Rāzī's commentary is typical of many others, both in its linguistic turn and in its emphasis on creation's benefits for humans as evidence of the existence of the willing creator. Notably, Al-Rāzī also denies natural determinism and insists on assigning agency exclusively to God. Commentaries often focus not just on the meaning and appropriateness of using certain terms but also on the logic of their order of appearance in the Qur'ān. Al-Rāzī, for example, explores the reasons why the word "heavens" occurs before "earth" in most mentions in the Qur'ān. Among the virtues of the heavens is that God ornamented them with the bright stars, the sun and the moon, and the throne, the pen, and the preserved tablet. God uses complimentary names to refer to the heavens to stress their high status. They are, after all, the abode of angels, a place where God is never disobeyed, the place toward which prayers are directed and toward which hands are raised in supplication, and they have perfect color and shape. The one advantage of the heavens over the earth that invokes a common scientific view of the time is the notion that the heavenly world influences the sublunar world, whereas the earth is not actor but acted upon. Al-Rāzī also lists some of the merits of the earth according to those who prefer it to the heavens, including its being the place where prophets are sent and mosques for the worship of God are built (1–2:106–7). Noticeably absent in this comparison is any discussion of a natural superiority of heaven over earth, the standard argument of natural philosophy, with which Al-Rāzī was fully familiar.[12] Rather than using the Qur'ān to elucidate science, or science to extract the proper meaning of the Qur'ānic text, the quasi-scientific discussions of exegesis often aim at explaining the order of words and at demonstrating the linguistic, rhetorical miracles of the Qur'ān. Indeed, the commentaries emphasize not just the creation of a perfect and wondrous world but also God's use of words to refer to his creation, in language that cannot be emulated by even the most eloquent humans (105).

The marvels of creation are a recurrent theme of Qur'ānic commentaries. They are viewed as signs of God and proofs that he exists, is all-powerful and all-knowing, and is the willing creator of all being. One of the commonly cited verses that urge contemplation of the signs of the heavens and the earth reads: "In the creation of the heavens and the earth, the alteration of night and day, are signs for the wise. Those who remember (pray to) God, standing or sitting or lying on their sides, who reflect on and contemplate the creation of the heavens and the earth, (say): Not in vain have You made them. All praise be to you, O Lord preserve us from the torment of Hell" (Qur'ān 3:190–91). Al-Rāzī, in his commentary on the verse, contends that the human mind is incapable of comprehending how a small leaf on a tree is created, how it is structured, or how it grows, which means that the larger task of discovering God's wisdom in the creation of the heavens and the earth is next to impossible. Humans must therefore concede that their creator is beyond full comprehension and admit his utmost wisdom, and the great secrets of creation, even if there is no way of knowing what these secrets are: when people reflect on the heavens and the earth, they will eventually realize that the heavens and the earth were not created in vain, but that their own "intellects are incapable of comprehending" such "remarkable wisdom and great secrets" (8–10:137–41). Therefore, the ultimate purpose of reflection is to establish the limitations of human knowledge and our inability to comprehend creation, not to establish a scientific fact and demonstrate its correspondence with the Qur'ān. The contemplation that the Qur'ānic text calls for is outside the text, in nature, and does not move back to the text, nor does it follow or correspond to any particular Qur'ānic scheme. The commentaries share this understanding of contemplation. Thus, contemplation does not imply a correlation between science—whether natural philosophy, astronomy, or medicine—and the Qur'ān. The Qur'ān, according to the commentaries, directs people to reflect on the wisdom of the creation of nature but provides no details on the natural order or on ways to decipher it. Such details, when they appear in classical Qur'ānic commentaries, are drawn from the prevalent scientific knowledge of the time. A brief overview of the way commentators invoke creation as evidence of God and his traits illustrates this fundamental divide between science and the Qur'ān.

As noted above, the Qur'ānic signs of creation are often classified into signs from within the self (*dalā'il al-anfus*) and signs from the external world (*dalā'il al-āfāq*). Alternatively, the signs are classified into signs in the heavens, on earth, or in what falls in between. Among the heavenly signs are the movements of the celestial orbs and their magnitudes and positions, as well as signs specific to different components of the heavens, such as the sun, the moon, and the planets. The earthly signs include minerals, plants, and humans.[13] Many commentaries do not dwell on the signs, but when they do, the most striking feature of the discussions, especially discussions of the heavenly signs, is the mixing of information drawn from astronomy and natural philosophy with a wealth of nonscientific information. For example, the benefits of the rising and setting of the moon are that its rising helps night travelers find their way, and its setting helps fugitives seeking escape and shelter from their enemies. Shooting stars, or meteors, drive devils away and keep them from spying on angels in the heavens (1–2:108–9).[14]

The benefits to humans are a common feature of the commentaries on the "sign verses," just as they are of commentaries discussing the creation. While the complexity and perfection of creation is, in and of itself, a sign of the wise creator, the primary proof of his wisdom is not the creation but the benefits to humanity that accrue from it. A typical commentary focuses on the specific way that various aspects of the natural phenomena are arranged to maximize the benefits they offer to humanity; since there is no inherent reason for the universe to be arranged in a particular fashion, there must be a willing maker who chose to create the universe the way it is, and the guiding principle of this creation must have been human benefit. That is, the manifest benefits to humans in the way things are ordered prove the existence of a wise and willing creator. To be sure, the subjugation by God of all creation in the service of human beings serves both their needs for survival and their independence, without which they could not worship God; the benefits are in both this world and the hereafter.[15] Regardless, benefit and utility were not the ultimate purposes of creation; rather, benefit is what induces people to reflect on God's creation, recognize the magnitude of his power, and believe in him.

Classical commentaries often introduce elaborate discussions of scientific subjects to illustrate the idea of God's wise choices in creation as ways of maximizing human benefit. Al-Rāzī, for example, in his commentary on Qur'ān 2:22, outlines the prerequisites for making the earth a bed (*firāsh*). After asserting that one of these prerequisites is that the earth not move, Al-Rāzī proves his contention (1–2:101 ff.). If it were to move, its motion would be either linear or circular. If it were linear, it would be falling; yet, since heavier objects move faster than slower ones, the earth would fall at a faster speed than people on its surface, and as a result, they would be separated from the surface of earth and could not use the earth as a bed. If, on the other hand, the earth's motion were circular, the benefit for humans would not be complete, since people moving in a direction opposite to its motion would never reach their destinations.

In pursuing this point, Al-Rāzī surveys the evidence adduced by various scholars to prove that the earth is stationary. What follows in his commentary is a quasi-scientific discussion that draws on but does not privilege science as the authoritative reference on this subject. Some scholars, Al-Rāzī reports, argue that the earth is bottomless and thus has no bottom to move to. This view, Al-Rāzī contends, on theological and not natural-scientific grounds, is wrong because all created bodies are finite. Others concede the finitude of objects, but argue that the earth is stationary because it is a hemisphere whose flat bottom floats on the surface of water. Al-Rāzī rejects this argument, too: even if this were true, both the earth and the water on which it floats could be moving. And, he wonders, why would one side of the earth be flat and the other round? Again, although Al-Rāzī could have invoked arguments for the sphericity of the earth that were more in line with the sciences of the time—and he was thoroughly familiar with them—his response here is notably general and not grounded in science.

Irrespective of how scientific these and other arguments appear to us today, these discussions did not reflect the prevalent scientific view of Al-Rāzī's time, either. The closest he got to engaging the prevalent understanding of science was when he reported, and rejected, the Aristotelian argument that the earth, by nature, seeks the center of the universe. This, Al-Rāzī rightly notes, is the view of Aristotle and the majority of his followers among the natural philosophers. Al-Rāzī objects to this view on the grounds that the earth shares the trait of physicality with all other bodies in the universe, and its acquisition of a specialized trait that would make it stationary is logically contingent. Thus, it is the free volition of the maker and not the inherent nature of the earth that accounts for its stillness. If anything, Al-Rāzī adds, the nature of the earth is to sink in water, and God reverses its nature so that it does not submerge, in order to maximize the benefit that the earth provides for humans and make it a place where they can reside (102–4).

This elaborate, quasi-scientific discourse that draws freely on the scientific knowledge of the time is evidently not aimed at upholding a particular scientific view of nature. Nor is it intended to make positive contributions to the accepted body of scientific knowledge. Rather, Al-Rāzī's primary purpose is to argue the contingency of created order and its ultimate dependence on God. Nowhere in this or other classical commentaries do we encounter the notion that a certain scientific fact or theory is predicted or even favored in the Qur'ān. Instead, the commentaries emphatically reject explanations of Qur'ānic verses that are grounded in the notion of a natural order. The sign verses serve as evidence of the creator, not in the particular knowledge that they convey about nature but in the ultimate conclusion of each and every verse: that a choice was made in creation, so there is a creator who made this choice. As Al-Rāzī puts it, the "world is created with perfect management, comprehensive determination, utter wisdom, and infinite omnipotence" (109).

Discussions of the natural phenomena in Al-Rāzī's commentaries conform with the general outlines of such discussions in other classical commentaries in two main respects. First, none of them use the Qur'ān as a source of knowledge about nature, and second, their authors usually undertake the exposition of various scientific theories and explanations not to favor one over the others but to suggest that there are multiple possible explanations, on all of which the Qur'ān is neutral.

Asserting the multiplicity of possible explanations of natural phenomena is hardly compatible with a positivist scientific outlook. Classical exegeses, however, are full of such assertions. Most commentaries on the sign verses offer multiple interpretations, only some of which are connected to science. Although some of the interpretations are rejected, many are allowed; and information culled from scientific discourse is often countered, rather than confirmed, by alternative interpretations that are considered acceptable. For example, Qur'ān commentators often maintain that the role of meteors is to preserve the heavens by driving away devils, or that the movement of the sun to a resting place (*tajrī li mustaqarrin lahā;* Qur'ān 36:38) refers to its movement to a point beneath the throne, where it prostrates itself, then rises again.

In a move that further clarifies his exegetical strategy, Al-

Rāzī notes in the commentary on this same verse (*tajrī li mustaqarrin lahā;* Qur'ān 36:38) that most commentators agree that the sky is a plane and has no edges or peaks (25–26:161–62). In response, he maintains, nothing in the text of the Qur'ān suggests that the sky has to be a plane and not a sphere. In fact, "sensory evidence indicates that the sky is actually spherical, so that must be accepted." After giving some sensory evidence to illustrate his point, he adds that such evidence is abundant, and its proper place is in astronomy books. To Al-Rāzī, therefore, just as in the discussions of the direction of the qibla, the authority on this matter is the science of astronomy, not the Qur'ān, however understood. The only reason for him to enter this extra-Qur'ānic discussion was to undermine the claims of commentators who wrongly extended the authority of the Qur'ān outside its proper realm.

Another aspect of Al-Rāzī's exegetical strategy with regard to the sign verses is revealed in his commentary on the same verse where he takes issue with astronomers, not commentators. Astronomers maintain that celestial orbs are solid spherical bodies, but Al-Rāzī contends that this is not necessarily the case. The basis for Al-Rāzī's objection is the possibility, astronomically speaking, to have an orb that is a circular plate or even to have an imaginary circle that a planet traces in its motion. It is not beyond God's power to create any configuration (25–26:76).[16]

Although Al-Rāzī's interest in these quasi-scientific subjects exceeds that of other commentators, scientific knowledge is freely invoked, as a general rule, and occasionally challenged in classical commentaries on the Qur'ān. Yet the purpose of rejecting some scientific views is not to promote alternative ones, nor to assert the authority of the Qur'ān at the expense of the various fields of science. In the absence of a clear statement in the Qur'ān, scientists seek answers to scientific questions in their respective fields. The contrary, however, is not true, since the text is not science. When a Qur'ānic text is in apparent conflict with a scientific fact, commentators do not present the Qur'ānic text as the authority. Rather, they explore the possibility of alternative scientific explanations, thus suggesting that scientific knowledge on such points of contention is not categorical.

It follows that in exegetical literature, religious knowledge and scientific knowledge are assigned to their own compartments. This separation would justify the pursuit of science, and even the use of scientific discourse in commenting on the Qur'ān, but would also limit the use of scientific discourse. A case in point is Al-Rāzī's contention that ignorant people may object to his unusual use of the science of astronomy to explain the book of God. But, he asserts, God has filled his book with proofs of his knowledge, power, and wisdom, which are inferred from the conditions of the heavens and the earth. If exploring these subjects and reflecting on them were not permissible, God would not have so frequently urged humans to reflect on these signs. "The science of astronomy," says Al-Rāzī, "has no other meaning than reflection on how God ordered the [heavens] and created its [different parts]" (13–14:121). The purpose of his exercise, he continues, is not to establish a correspondence between scientific verities and the Qur'ān but to reflect and hence reinforce belief in the creator of the impeccable universe. This kind of reflection to support belief does not produce knowledge about the natural order.

Despite all his talk about the permissibility of using astronomy in exegesis, Al-Rāzī asserts that all creation is from God, that the planets have no influence on the sublunar world, and that the "assertion of natures, intellects, and souls in the manner advocated by philosophers and diviners is invalid" (122–23). These statements are directed primarily at fellow religious scholars and not at scientists. When discussing the religious import of the Qur'ān, commentators are urged to stay with the text and not try to impose astronomical knowledge on it or, for that matter, feign a Qur'ānic understanding of astronomy. The Qur'ānic text to which Rāzī wants to restrict himself and his fellow commentators does not have scientific import, nor does it provide binding scientific facts. While it underscores the wisdom and power behind creation, it says nothing about the exact order and workings of the created world. The complexity and wondrous nature of the world reinforce belief in God, but belief is not contingent on the adoption of any particular scientific view. In fact, scientific facts and theories in themselves do not provide evidence of the oneness of the creator. The opposite is true: that other natural orders are possible points to a willing maker who chooses one of many possibilities.[17] According to this logic, everything in nature, however it is explained, as well as all scientific discoveries and facts, irrespective of their certainty, would serve as proofs of the existence of the maker. And this is the fundamental reason why the scientific and the unscientific appear side by side in the commentaries on the Qur'ān.[18]

Kalām

The genre of Islamic religious scholarship that comes closest to philosophy is, without question, kalām (speculative theology), not only because reason and religious tradition intersect in it but also because the genre was shaped by deliberately drawing on the Greek philosophical tradition.[19] In regard to natural science in particular, some early works of speculative theology included substantial discussions of physics. In studying physical theory in early speculative theology, a contemporary scholar, Alnoor Dhanani, illustrates how significant aspects of physical theory—theories of matter, space, time, and motion in particular—were discussed.[20]

In contrast to the earlier works of speculative theology, the scientific focus of later works shifts to astronomy instead of physics. Over time, these works were increasingly comprehensive. In the formative period of the Islamic scientific tradition, the first rational sciences were practical sciences, such as mathematics, algebra, geometry, medicine, the manufacture of automata, astronomy, and—this was added gradually—logic. The early specialized scientific works did not treat doctrinal matters and were not employed doctrinally, nor were they given metaphysical or mystical interpretations.[21] Treated individually, they did not compete with the religious sciences. However, shortly after their fragmented introduction, their philosophical foundation was recognized—or discovered—with the works of such philosophers as Al-Kindī and Al-Fārābī, and so were the implications of the philosophical framework in metaphysics and astrology. In these areas, the philosophical sciences competed directly with religion. Nonetheless, the first full-fledged philosopher, Al-Kindī, was also a full-fledged scientist who composed specialized works in medicine, optics, astronomy, and mathematics. Al-Kindī also employed philosophy doctrinally in defense of Islam. With Al-Fārābī, however, the coherence of the philosophical doctrine was reestablished in its own right, and philosophy was no longer subservient to religious doctrine. Al-Fārābī composed works on mathematics and logic, but he was primarily a philosopher, and even though he discussed religious matters, he did so on philosophy's terms.

A notable trend among philosophers from Al-Fārābī's time on was their increasing specialization in philosophy proper as distinct from the mathematical sciences. Ibn Sīnā excelled in both philosophy and medicine, of course, but the mathematical part of his philosophical encyclopedia was not of the same caliber as the other parts, nor did it compare to original research in the mathematical disciplines at his time. A century and a half after Ibn Sīnā, Ibn Rushd distinguished himself as a specialized philosopher and the leading commentator on Aristotle, although his other scientific works (for example, on medicine and astronomy) evinced definite regression in their respective fields.[22] Overall, the specialized scientific education of philosophers in the field of mathematics did not seem to go beyond the basics of Greek mathematics and did not benefit from the numerous additions of the Arabic mathematical tradition.

The exact opposite trend seemed to be at work in the case of the theologians—the *mutakallimūn*, or scholars of kalām. Whereas the earliest works of speculative theology were, in effect, inventories of doctrinal positions,[23] these works quickly gained in complexity and systematization. With the passage of time, the doctrinal aspects of these works decreased relative to the amount of specialized scientific knowledge incorporated into them. In fact, many of the great mathematical scientists of the later periods were also religious scholars who composed theological works as well as mathematical or astronomical works.[24] Given the increased incorporation of philosophical and astronomical discussions into theological works, some historians have argued that speculative theology came to replace Aristotelian philosophy as the quintessential Islamic philosophy.[25] There is no doubt that theological works became more coherent and systematic with the passage of time, but, as I will argue below, they were not aimed at providing a comprehensive alternative to Greek philosophy, nor did they claim to be a comprehensive system of thought that underlay all other sciences.

One of the most influential works on theology even today is Aḍuḍ al-Dīn al-Ijī's (d. 1355) *Kitāb al-Mawāqif fī 'Ilm al-Kalām.*

Al-Ijī wrote extensively on astronomical subjects, but he did not write a separate work on astronomy, and his comments in *Al-Mawāqif* were not meant as independent contributions to astronomical knowledge. Rather, Al-Ijī's interest was theology, and he discussed astronomy only to the extent that it was connected to theological problems. In a study of Al-Ijī's work, A. I. Sabra highlights the emphasis in *Al-Mawāqif* (and in the works of all theologians, whether Mu'atizilīs or Ash'arīs) on speculative theology as a genuine form of knowledge.[26] Sabra also sets forth the definitions, questions, and subject matter of speculative theology that Al-Ijī provides: "What is ultimately established in *kalām* inquiry is . . . *not something given from the start*, but a distinct product of the inquiry itself."[27] As Al-Ijī puts it: "All (religious) sciences draw from *kalām*, whereas it draws from none. It is absolutely at the head of all (religious) science." Sabra contends that this was "not a confused *kalām* but a confident *kalām* on the offensive against *falsafa*," and that "Muslim intellectuals at large . . . naturally regarded the *mutakillamūn* as the true thinkers of Islam, those who struggled from an Islamic perspective to forge a truly Islamic philosophy."[28] This kind of theology, says Sabra, can rightly be designated "philosophical theology," since it "produced a theory of knowledge, a cosmology (or theory of the world—the *falāsifa*'s 'world of nature'), and a metaphysics (a theory of transcendent being), all of which were presented by theoretical arguments expressed in abstract terms."[29]

The image that emerges is of theology as an autonomous and "truly Islamic" philosophical system that exercises hegemony, or wishes to, over all other intellectual disciplines, but subjects itself to none. But I would like to propose a different reading of Al-Ijī. In his definition of *kalām*, Al-Ijī maintains that it is "a science through which it is possible to confirm (*ithbāt*) religious dogma either by means of providing proofs or by [means of] removing ambiguities."[30] Al-Ijī adds that even an adversary with whom he may disagree may still count as a scholar of kalām.[31] Clearly, therefore, the truth of kalām, according to Al-Ijī, is not an absolute truth. Al-Ijī also maintains that the questions (*masā'il*) of kalām include "every theoretical judgment (*ḥukm naẓarī*) concerning a knowable thing that is either one of the religious beliefs or one [a judgment] on which the establishment of a religious belief depends."[32] To be sure, Al-Ijī maintains that kalām is the most noble discipline, but this is because its subject matter, the divine and related matters, is noble and not because theological knowledge has a privileged status in fields other than theology. The notion that other sciences depend on kalām, whereas it depends on none, also needs some qualification. What Al-Ijī actually says is that kalām does not depend on the evidence of other derived disciplines, but its starting point remains transmitted evidence (*dalīl al-naql*), and it sets out to prove the religious dogma that has already been established through transmission.

Perhaps a more important issue here is whether kalām represented a complete and purely Islamic philosophical system. Sabra contrasts this notion of kalām as a philosophical system and a genuine intellectual pursuit with the common understanding of kalām as an apologetic exercise.[33] It is possible, however, to conceive of kalām as an apologetic undertaking that is not a complete philosophical system without diminishing its value as a genuine intellectual pursuit. In fact, this is how Al-Ijī presents kalām. For example, in a discussion of the reasons for which the word *kalām* is used to designate the field of theology, Al-Ijī mentions three possibilities: that the field was called kalām because it involved disputation with philosophy; or because it was named after the most controversial question discussed in it, the speech (kalām) of God; or because it provides a specialist in the field with the ability to engage in debates over religious subjects and to respond to opponents.[34] All three possibilities involve disputation as a central undertaking. Above all, however, kalām, as conceived by Al-Ijī, was not a complete philosophical system, and its scope of intellectual authority was certainly not comparable to the scope claimed for philosophy proper. In fact, despite all his professions that theology is great, Al-Ijī's ultimate purpose was to undermine the comprehensive claims of philosophy as a complete system of knowledge, not create an alternative comprehensive system. This is what underlies Al-Ijī's recurrent assertions that it is possible to imagine explanations of various natural phenomena that are different from the ones presented by philosophers as definite and conclusive explanations.

The field of theology for which Al-Ijī claimed authority in religious matters had no other epistemological claims outside the field itself. In Al-Ijī's expositions of various astronomical theories or theories of natural philosophy, which occupy a substantial part of his book, a standard formula recurs—"Why is it not possible?" (*limā lā yajūz*)—to posit an alternative explanation.[35] But Al-Ijī himself never advocates specific alternatives. In concluding his extensive discussion of astronomical problems, he maintains that the orbs and circles posited in astronomical models are all "imaginary things (*umūr mawhūma*) that have no external existence (*wa lā wujūd lahā fī al-khārij*). (Religious) prohibition does not extend to them, and they are neither objects of belief (*la ḥajara fī mithlihā wa la tat'allaq bi-i'tiqād*), nor subjects of affirmation or negation (*la yatwajjah naḥwahā ithbāt wa ibṭāl*)."[36] In other words, these matters do not belong to theology, and kalām has nothing to say in their regard. Al-Ijī mentions them just to undermine philosophers' inclusion of them in theological arguments. Al-Ijī also objects to the attribution of various astronomical phenomena to the volition of the heavenly objects—he reserves final agency to God (*al-Fāʿil al-Mukhtār*).[37] In his words, "the purpose of reporting these differences is . . . so that a rational and smart person may confirm that they have no categorical proof (ithbāt) for what they claim."[38] Rather than exerting the hegemony of kalām over these other disciplines, Al-Ijī tries to disengage these disciplines and to guard the autonomy and separateness of the specialized fields of knowledge. For him, religion does not provide principles for the sciences, nor does it play a selective role in the evaluation of rival scientific theories.

In fact, despite his strong polemical stance, Al-Ijī recognizes in several places the authority of the fragmented internal knowledge that he excludes from theology. In one instance, Al-Ijī uses a geometrical hypothesis (*fard*) to prove the finitude of distances.[39] He responds to the possible objection to this proof by saying that "reason asserts the possibility of certain hypotheses, such as a geometrical hypothesis; one can only reject its (validity) out of stubbornness."[40] Of course, Al-Ijī's discussion of these subjects was primarily for the polemical defense of kalām against its intellectual rivals, and he did not attempt to reconstruct or make positive contributions to such fields of knowledge as astronomy. As a result, many of Al-Ijī's conceptualizations of other sciences remain unclear. What is implied in Al-Ijī's work, however, is forcefully spelled out by other scholars, Ṣadr al-Sharīʿa al-Bukhārī for one.[41]

Ṣadr wrote a three-volume work on logic, theology, and astronomy, *Taʿdīl al-ʿUlūm* (The Adjustment of the Sciences), with each volume approaching its subject from a significantly different vantage point. In the volume on logic, Ṣadr responds to the widely accepted view on the use of logic: since the perfection of the soul is attained through the acquisition of theoretical knowledge, and since theoretical reflection is susceptible to error, one needs to establish a standard that safeguards from error (*qānūn ʿāṣim*), and this standard is logic. Ṣadr notes that logic deals only with formal questions (*fī al-ṣūra*) and hence cannot establish the validity of the basic premises to which it is applied (*al-qadāyā allatī hiya mawād al-aqyisa;* B12v). So error can occur at this substantive level as well through the erroneous use of logic. But Ṣadr devotes this volume to a discussion of the formal aspects of logic and leaves the substantive discussions to the volume on kalām.

The subject of kalām, Ṣadr argues, is the essence (*dhāt*) of God, including his existence and oneness, his attributes, and the conditions of existing things to which God gives essence (māhiyya) and existence (*wujūd;* 113v, 120v). Therefore, the discussion of essence and existence, of substances (*jawāhir*) and accidents (*aʿrād*), and of astronomical matters is simply on account of their dependence on God for existence. Another subject that belongs to kalām is discussion of ways to win the gifts that God promised believers on the Day of Judgment, and the explanation that these gifts cannot be attained without worship, which, in turn, is not possible without seeking the help and guidance transmitted by messengers, prophets, caliphs, and saints (*ahl al-risāla wal nubuwwa wal-khilāfa wal-wilāya*). Right at the beginning, therefore, Ṣadr asserts the authority of transmitted religious knowledge in kalām.

In the following discussion of such things as existence (wujūd) and being (*kawn*), Ṣadr argues that these are concepts that gain meaning only when they are in construct with something that has outside material existence (*ashkhāṣ;* 120v). The ultimate purpose of these and other discussions is to reject the notion of a conceptual existence (*wujūd dhihnī*, "universal concept") and to establish the dependence of all these conceptual entities on outside material existence, which emanates from God (121r–v). To be sure, Ṣadr uses sophisticated rational arguments in his expositions; he also endorses what he calls *ḥikma*, a sanctioned use of reason coupled with practice, and argues that it is different from philosophy (154r–v). However, the rational exercise undertaken to establish the dependence of all existence on God stops short of grasping the truth about God (*tanazzahat ḥaqīqat al-ḥaqq ʿan an tudrak ʿaqlan;* 134v). And although Ṣadr retains a role for reason, he clearly rejects the philosophers' use of reason as an absolute criterion for judging not just humans but also God (153r–156r). In short, Ṣadr's theology, like Al-Ijī's, is not meant as a complete philosophical system, and it uses rational arguments to limit rather than extend the overlapping of theological knowledge with other kinds of knowledge.

Ṣadr's recognition of the autonomy of astronomical knowledge is clearly illustrated in his volume on astronomy, where he ignores the metaphysical aspects of astronomy and focuses instead on the mathematical models representing the motions of the planets. In fact, in the volume on theology, Ṣadr sets the stage for the exclusion of all metaphysical principles governing the old astronomy. Against the established principles of Aristotelian natural philosophy, for example, Ṣadr argues that a void is possible (173v–174r); that universes other than our universe may exist (176v); that orbs as well as simple bodies are not by natural necessity circular (177r–v); that the motions of the heavenly orbs are neither natural nor voluntary, but compulsory (178r–v); that, based on most of what we know about the heavenly orbs, their nature is not simple but composite (179v); and that orbs do not have souls (184r–v). To be sure, Ṣadr's astronomy was not completely stripped of regulating principles, and one of the main objectives of Ṣadr's volume on astronomy was to produce mathematical models that did not violate these principles. Primary among these was a principle discussed in the previous chapter: that a uniform circular motion around an axis that does not pass through the center of the rotating sphere is a geometrical impossibility. After undermining metaphysical principles in his volume on kalām, the only rules left for governing Ṣadr's adjusted astronomy were mathematical. In other words, Ṣadr's work on kalām did not produce theological principles that would inform the new astronomy; it simply eliminated many of the principles of the old theology. In fact, the influence of astronomy on kalām seems to be more significant than the reverse. Such, for example, was the case with the various astronomical proofs that Ṣadr provides for the indivisible part (*al-juz' alladī lā yatajazza'*; 157v–162v).

It seems reasonable to suggest that the objective of Ṣadr's project of adjusting the sciences was to recast the sciences to allow for their separation and disengagement. This adjustment was informed by previous developments in the field of astronomy, most notably its systematic mathematization, and by the relentless questioning of its theoretical foundations. Many of the objections to natural philosophy expressed by Al-Bīrūnī are echoed in Ṣadr's work. The new science, therefore, was not an interpretation of an old science in light of a new theology.[42] Rather, the science emerged long before, and was largely responsible for, the production of the adjusted kalām.

A main trend in the discipline of speculative theology, then, as in exegesis, was to disengage the specialized sciences from their original philosophical framework and from the metaphysical, thereby allowing them to be restructured free of the demands of natural philosophy, metaphysics, and theology. The new works on kalām did not offer an alternative metaphysics to guide and bind science. Rather than being a source of knowledge on physical theory, later works on kalām were themselves informed by developments in the field of astronomy; as a result, the scope of theology was redefined and transformed from an encompassing system of knowledge into one with limited epistemic authority outside the area of metaphysics.

The Problem of Causality

Almost invariably, discussions of the Islamic attitude toward science invoke the works of Al-Ghazālī, whose views on the various sciences have received more scholarly attention than any by other Muslim scholars who wrote on the subject. The debate on his true attitudes and views continues among contemporary scholars, who have not reached consensus even over the interpretation of his most explicit work, *Tahāfut al-Falāsifa* (The Incoherence of the Philosophers),[43] let alone an integrated assessment of his oeuvre.[44]

Numerous historians have focused on Al-Ghazālī's denial of causality and argued that by repudiating a cornerstone of natural science, Al-Ghazālī undermined the very possibility of scientific knowledge. So let us look more closely at Al-Ghazālī's view on this subject. In his *Tahāfut al-Falāsifa,* he delivers a systematic criticism of what he considers to be the flaws and shortcoming of philosophy in matters related to metaphysics.[45] Al-Ghazālī maintains that his differences with philosophers fall into three categories: linguistic conventions, which he chooses not to discuss; matters that do not involve the denial of a religious principle; and philosophical assertions that contradict the dictates of religion. Believing in the doctrines of messengers and prophets, Al-Ghazālī adds, does not require opposition to philosophers (77–78).[46] "Someone who thinks that denying these matters is part of religion commits an aggression against religion and weakens it, because geometrical and mathematical proofs over which there is no doubt are based on these matters. Thus, someone who learns these things and verifies their certainty . . . would only doubt religion and not this [science] if he is told that the two are in opposition" (78).

Al-Ghazālī thus asserts that part of the philosophical sciences is not outside the scope of religion. A critic, Al-Ghazālī says, might invoke the tradition of the Prophet that says, "The sun and the moon are two of the signs of God; they are not eclipsed for anyone's death or life, so if you see an eclipse, rush to mention God and pray." Al-Ghazālī's response is that nothing in this tradition subverts the natural explanation of the eclipse; the command to pray at eclipses, like the command to pray at sunset and sunrise, is a conventional religious obligation that has nothing to do with the natural phenomenon of the eclipse. Even if there is a tradition that contradicts scientific knowledge, Al-Ghazālī says that it would be far better to interpret it than to place it in opposition to certain knowledge. Nothing pleases the enemies of religion more than a statement by their rivals asserting that certain knowledge is in opposition to religion (79).

The "inquiry into the natural world" that concerns Al-Ghazālī is "whether it is eternal or created. Once it is determined that it is created, it does not really matter whether it is a sphere or a plane or a hexagon or a pentagon, or whether the heaven and what is below it is in thirteen layers, as they claim, or more or less than that. Assigning the inquiry into this matter to metaphysics is like assigning to it an inquiry into the layers of an onion or the number of the seeds in a pomegranate. The only thing that is relevant [to metaphysics] is that it [the world] is created by God, irrespective of its form" (79, 82). In this manner Al-Ghazālī expresses his desire to disengage natural science, or at least the empirical part of it, as well as mathematics, from metaphysics. The only part that ought to be opposed and undermined is that which relates to a conflict over one of the principles of religion, such as the belief that the world is created, the attributes of the creator, or the resurrection of bodies.

Al-Ghazālī takes the argument a step further: not only are the mathematical sciences and most of physics epistemologically distinct from metaphysics, but metaphysics does not rely on these sciences in its assertions. The only science deployed in metaphysics is logic. Philosophers do not monopolize use of this science or instrument, however; it is shared by all the sciences, though under different names and with different terminology (83). In an interesting section of his book, Al-Ghazālī rejects the use of arguments from mathematical astronomy to prove the eternity of the world by invoking the uniform motion of the heavenly orbs. Al-Ghazālī maintains that all the motions that can be generated by a combination of the motion of the celestial orb from east to west and the motion of the orb below it in the opposite direction can be generated by reversing these two motions, as long as the difference between the two motions is the same in reversal as it was before (103–4). Elsewhere Ghazālī responds to the organic cosmological argument that heaven is an animal and has a soul and is moved through the volition of this soul. In a response that captures his views and those of the other scientists and religious scholars we have considered so far, Al-Ghazālī avers that he does not claim the logical impossibility of a living celestial orb, since neither the large size nor the circular shape of the celestial orb prevents it from having a life, and since it is certainly within God's power to create life in any solid. However, Al-Ghazālī adds, "What we deny is their [the philosophers'] ability to verify this through a rational proof." According to Al-Ghazālī, therefore, the philosophical proof for the organic, living cosmos is not certain; at best, philosophers can prove the possibility, not the actuality, of its life (216–19).

Al-Ghazālī raises similar questions about natural science (physics). After listing its principles and branches, he states that religion is not in opposition to most of its subfields. There is conflict over only four assertions: that there is a necessary connection between physical causes and effects; that human souls are self-sufficient entities that are not imprinted in bodies, and that death simply means the departure of these souls from the body; that souls do not perish; and that souls cannot be returned to bodies (234 f.). Of these objections, the suggestion that souls do not have an existence independent of bodies seems to provide a rather material or "scientific" explanation of life. Once again, Al-Ghazālī does not rule out the possibility that the soul has a separate existence; he simply argues that the natural philosophical proof of its existence is not certain, nor can the matter be resolved through rational proofs, because its subject matter falls outside the material domain of reason (254). The main item on Al-Ghazālī's list of objections, however, is the issue of causality.

Al-Ghazālī attributes causality to habit and concomitance. For example, from our observation of the recurrence of events, and from our observation of the concomitance of two events, we have come to expect that when a piece of cotton is touched with fire, it burns. The expectation of burning is not established through demonstrative proof. Logically, all we can demonstrate is that burning occurs *when* and not *because* fire and cotton are brought together. According to Al-Ghazālī, the causal connection is a mental assumption that we make in our minds, not a relationship between the concomitant natural phenomena; that is, the laws of causality do not apply to nature but to our knowledge of it and to the nature of the perception that God creates in humans (242–43). Basically, Al-Ghazālī is questioning the deterministic connection between cause and effect in nature while asserting that the only possible knowledge about nature is predicated on a perception of a causal nexus. It follows that it is possible to organize knowledge about nature in a science that assumes causality, and that it is possible to order our (partial) perception about the natural phenomena in a system of knowledge. According to Al-Ghazālī, since God's ultimate causes are not knowable, all knowledge about the natural world is contingent and not certain. However, this knowledge is derivable from the world itself.

Al-Ghazālī's notion of the habitual order of God's continuous creation, like his denial of causality, are grounded in the Ash'arī school's doctrine of atomism.[47] According to this doctrine, God re-creates the world in all of its details in successive discrete moments, so what we conceive of as a continuous natural order is in fact a series of independent moments created separately by God. As such, the natural order we observe can be disrupted at will by God, and this disruption explains the occurrence of miracles. On the other hand, the habitual order of God's continuous creation produces in us the knowledge of a habitual natural order that describes the manner in which God chooses to act in the world. The result is a natural order in our minds that is not binding on God and yet is bound by the natural world as we perceive it.[48]

Al-Ghazālī's views on the various sciences were quite influential, and in some cases, there is evidence of the direct effects of these views on the development of some scientific disciplines. His discourse apparently helped naturalize logic and provide Islamic justifications for it. Such was the assessment of the famous fourteenth-century historian Ibn Khaldūn, who remarks in his *Al-Muqaddima* that after Al-Ghazālī, all religious scholars studied logic, but they studied it from new sources, such as the works of Ibn al-Khaṭīb (d. 1374) and Al-Khūnjī (thirteenth century), and that people stopped using the books of the ancients. "The books and the methods of the ancients," says Ibn Khaldūn, "are avoided, as if they had never been." Later he adds, "It should be known that the early Muslims and the early speculative theologians greatly disapproved of the study of this discipline [logic]. They vehemently attacked it and warned against it. . . . Later on, ever since al-Ghazālī and the Imām Ibn al-Khaṭīb, scholars have been somewhat more lenient in this respect. Since that time, they have gone on studying (logic)."[49] Yet, of Al-Ghazālī's views, and those of Ash'arite scholars in general, the aspect that had the most influence on the development of science was the concept of multiple possibilities (*tajwīz*), the notion that specific natural philosophical explanations (or planetary models) are possible but not certain, and that there may exist alternative explanations for the natural phenomena. A similar attitude is detectible in tafsīr and a variety of theological works and is echoed in the works of scientists from Al-Bīrūnī all the way to Al-Khafrī. As we have seen, this idea was grounded in an epistemological criticism of Aristotelian metaphysics, but it did not impose an "Islamic" metaphysics in its place.[50] In other words, Islamic theology helped undermine the metaphysical foundations of the old sciences without claiming a role for Islamic doctrine in the cognitive development of these sciences.

The call to disengage science and metaphysics was eloquently articulated by Ibn Khaldūn. Ibn Khaldūn recognized the possibility of occult knowledge (magic, talismans, numerology, alchemy, astrology), but in six chapters of his *Al-Muqaddima* he provided a thorough critique of occult knowledge on both ethical and epistemological grounds. While arguing that some effects from these forms of knowledge might materialize, he insisted that knowledge of these effects does not come through the rational faculties. In other words, these occult sciences are not subject to the rational controls that regulate the practical sciences, so they are not subject to the regulatory processes that order civilization (*'umrān*).[51] On similar grounds, Ibn Khaldūn provides a critique of metaphysics and divine philosophy (ilāhiyyāt). "The ultimate subject" of metaphysics, he argues, "is utterly unknown and cannot be reached by or be the subject of demonstrative proofs, because the abstraction of intelligibles (ma'qūlāt) from individual existents that have physical reality is possible only for things that we can perceive (*mudrak lanā*), and we cannot perceive the spiritual essences (*al-dhawāt al-rūḥāniyya*) in order to abstract other essences from them."[52]

In other words, Ibn Khaldūn does not deny the existence of spiritual essences, but he insists that they are outside the realm of a practical, civilizational system of knowledge (*'ilm ṣinā'ī* or *'umrānī*). And because the conditions of theoretical demonstrative proof do not obtain in these subjects, technically speaking, these practices do not fall within the rational sciences. Ibn Khaldūn does distinguish between the occult sciences (*'ulūm ghaybiyya*), which he criticizes on moral, methodological, and epistemological grounds, and metaphysics (ilāhiyyāt), which he sees as epistemologically deficient, though employing methods that can be beneficial. "Even if these methods are not suitable for their objectives [knowledge of matters related to the divine], they continue to provide the soundest rules of inference (*qawānīn al-anẓār*) available to us."[53] While rational intellection does not apply to the realm of God, the rules of rational reflection in themselves are sound. On these grounds, Ibn Khaldūn accepts logic, which, he argues, became legitimate and widely accepted after Al-Ghazālī's defense of it.[54]

As noted earlier, Ibn Khaldūn is not making an apologetic religious critique of a competing philosophical system, for he uses the same arguments to undermine the claim by philosophers that they can provide demonstrative proofs (burhān, *dalīl ʿaqlī*) for the logical necessity of prophethood or divine law as a natural need of human association and social order (*ijtimāʿ*). Ibn Khaldūn maintains that history is full of examples of forms of social order that were not predicated on the existence of prophets or divinely inspired legal codes. If anything, prophets and divine law are special cases of the general rule in that they provide a special, and certainly desirable from Ibn Khaldūn's perspective, form of group solidarity and coercive power, which are the essential ingredients of any social order.[55]

Now to sum up. As I have frequently noted, the contexts and conceptions of the relationship between science, philosophy, and religion in classical Islamic culture were extremely diverse. This makes it hard to characterize the relationship with any finality. Yet I provided evidence for one important epistemological approach that characterized the practice of science in classical Muslim societies, an approach that emerged as a conscious alternative to Greek philosophical epistemologies. Greek philosophy continued to thrive and evolve, but many of the most important developments in the sciences occurred in the context of the new epistemology.

The first important consequence of the new epistemology was to divest scientific knowledge of religious meaning (notwithstanding the marvels of God's creation). The pursuit of natural knowledge was no longer subordinated to metaphysical concerns, and as a result, it was also no longer subordinated to theological concerns. After Al-Ghazālī, the need to invoke religion to vindicate science considerably decreased, not because science was not accepted but because it did not need vindication. Excluding final-cause explorations from science did not compromise the providence of God, which was simply assumed without questioning (*bilā kayf*). When we read Al-Ijī alongside Al-Ghazālī and works of exegesis, it becomes abundantly clear that diminishing the authority of the Qur'ān in matters of natural philosophy was not a concern for these thinkers, because its authority, as Al-Bīrūnī asserts, was never assumed to start with. Therefore, the exclusion of first-cause explorations from science simply alleviated the burden of metaphysics and freed scientists to pursue the knowable. The ultimate meaning of any particular science was no longer sought by necessity in a higher philosophical or religious truth; instead it was sought within the distinct sphere of autonomous scientific knowledge. With this new epistemological posture, scholars set forth, clearing the path to the qibla of Fes (with which I opened the discussion in the first chapter). Now science could boldly claim authority based on secular reason even in the sacred space of religious ritual.

To be sure, scientific autonomy came at a price. For better or for worse, science did not turn into an ethical value to which other values would be subordinated. Yet while religion dominated the moral sphere and claimed a higher rank there on account of the nobility of its subject matter, it did not exercise an epistemological hegemony over science. That the unity of knowledge would give meaning to a hierarchical ranking of systems of knowledge was no longer assumed (irrespective of whether science or religion comes first).[56]

The second major consequence of the new epistemology was related to this lack of an epistemological hierarchy: science was conceived as a culturally neutral or universal activity. The ideal of cultural neutrality is once again eloquently captured by Ibn Khaldūn, who refers to the rational sciences as sciences shared among all the nations: "The intellectual sciences are natural to man, inasmuch as he is a thinking being. They are not restricted to any particular religious group. They are studied by the people of all religious groups, who are all equally qualified to learn them and to do research in them. They have existed (and been known) to the human species since civilization had its beginnings in the world."[57] With this ideal of neutrality, it was possible to pursue the sciences for their intellectual merits independent of any connection to religion.

Chapter 4

In the Shadow of Modernity

The title of this book is *Islam, Science, and the Challenge of History*. My initial challenge was to write a justifiable and nuanced history of Islamic science. Here the questions were how to understand the sciences in classical Muslim societies in their historical contexts; how to avoid the temptation of writing the history of Islamic science as a history of scientific precursors and forerunners; how to provide a reasonably accurate account of the continuously changing social contexts of scientific production; how to achieve a deep, multilayered understanding of the conditions of production and reproduction of a scientific culture; how to account for the internal epistemological coherence of the Islamic scientific tradition; and how to situate the sciences in their cultural context and in relation to other cultural forces.[1] My objective in the first three chapters, then, was to provide an informed understanding of the Islamic culture of science as one of the constitutive forces of classical Islamic culture.

In this fourth and last chapter, the problematics are of a different nature. The challenge for the historian remains, although this time it is the challenge of thinking about science in Muslim societies in the absence of a living scientific culture. But there is a second, perhaps more existential challenge that a reasonably grounded account of the history of Islamic science poses to contemporary understandings—understandings that, I should add, often fly in the face of the actual historical experiences in classical Muslim societies.

This chapter could have easily been entitled "Islamic Science after the Fall." My objective here is to explain absences, a task far less interesting and much more speculative than trying to explain things that happened. The main problem is that the Arabo-Islamic scientific culture is a legacy of the past and a hope for the future but absent, in effect, in the present. Much of the debate about science in the modern period focuses on bridging the gap between the lost past and the desired future. Of course, the proposed modes of bridging this gap depend on how the past is imagined, and as we have seen, much remains to be settled on this front.

In the previous chapters I provided a historical account of some aspects of Arabo-Islamic scientific thought, its epistemological foundations, its cognitive tools, and its position within the larger Islamic culture. I offered an interpretation of this past based on readings of what I think are representative texts and practices, and not on today's metaphysical assumptions and desires. There is nothing unusual or inherently wrong about the dependence of modern interpretations on new constructions of historical narratives, which often involve significant departures from traditional historical narratives. In a sense, all histories have the present as their point of departure, and I do not wish to undermine modern readings and historical interpretations because of their presentness. The point I am trying to make is that although historical facts (and scientific facts) are not independent of the observer's vantage point, they ought not to be the mere haphazard imaginings of the historian.

I would suggest, therefore, that perhaps one of the tasks of the historical investigation of science in modern Muslim societies is to explain the needs and conditions that give rise to modern Islamic discourses about science. These conditions include a variety of moral imperatives, as well as the need to redeem the self and make up for conceived deficiencies and the need to construct ideals and imagine scenarios for overcoming the current predicaments of underdevelopment. Bridging the gap also depends on ways of imagining the desired future. But above all it depends on formulating and executing complex developmental plans in the context of the social, economic, and political constraints of the present. With the exception of occasional impressionistic remarks, most of these issues fall outside the scope of the present book.

On Decline

Before I move to the modern period, with its manifest deficiencies, a word about "decline" is in order. In the previous chapters I deliberately avoided approaching Islamic science from the perspective of its eventual decline. To my mind, attempts to interpret eight centuries of vibrant scientific activities in light of the eventual decline of Islamic scientific culture (or its failure to produce a scientific revolution) are inherently problematic.[2] The fundamental assumption of histories of decline is that the eventual waning of scientific activity resulted from the intrinsic, culturally predetermined disposition of the Islamic sciences, not from specific historical developments. In other words, even when Islamic scientific culture was flourishing, the seeds of its eventual fall were embedded within it.[3] In contrast to those taking this approach, I do not see a cultural or epistemological imperative for decline; rather, I view the decline as a product of specific historical developments, a symptom of history and not its cause. As such, an inquiry into the causes of decline falls outside the scope of the present study, which focuses primarily on the epistemological assumptions of Islamic science.[4]

Once we accept that the scientific decline was a historical occurrence and not an essentialized cultural determinant, it follows that the first step toward explanation is to settle the question of periodization. Rather than constructing a historical account based on the presupposed causes of decline, a periodization scheme derived from the actual historical record should inform the interpretive accounts of the rise and fall of science.[5] It also follows that the decline happened, when and if it did, in various places at different times, which means that there are probably different historical causes for it. In fact, numerous historical records talk about aspects of decline that were specific to particular regions and did not apply to all the Muslim world; additionally, a researcher gets the sense that the decline was reversible, as was actually the case in many instances. Historical sources for North Africa, for example, talk about the decline of the sciences and other intellectual activities in their regions while often noting that things were different in the Muslim east. Endemic factional wars, military confrontations with expanding European powers, and plagues are often cited as primary causes of intellectual stagnation. An Andalusian, Lisān al-Dīn Ibn al-Khaṭīb (d. 1374), in a famous will to his sons, says that the western part of the Muslim world is so troubled and destabilized by war that it has became "good for nothing but *jihād*."[6] Ibn Khaldūn describes the devastating plague (*al-ṭāʿūn al-jārif*) of the mid-fourteenth century that turned conditions in the Maghrib upside down, "decimated a whole generation and folded and erased many of the signs of civilization . . . the sharp reduction in the [number] of people led to the destruction of cities and livelihoods . . . as if the voice of the universe called on the world to stagnate and contract and it responded."[7] And yet, as in the case of Ibn Khaldūn himself, a large number of highly original works continued to be composed despite political and social setbacks.[8]

The setbacks described by Ibn Khaldūn were not generalized to all of the Muslim world, nor did they contribute to an irreversible decline. A few centuries later, however, other factors had long-term impacts that were harder to reverse. Without degenerating to economic essentialism, let me stress the economic context of such long-term decline. Many historians have noted, among other things, the gradual decrease in the maritime power of Muslim states relative to their European competitors, and the effects of this decrease on trade. But, as many studies have shown, the Ottoman economy was far from stagnant, and despite the relative decline in Ottoman maritime power and trade, the Ottoman land-based economy retained its vibrancy and often compensated for losses in maritime trade. More important, in the late Middle Ages, the price that a rising but relatively underdeveloped state such as Portugal was willing to pay in order to reach the Indian Ocean via the Cape of Good Hope was not one that the affluent and civilized Ottoman society was willing to pay, especially when land trade to India was open to the Ottomans. In other words, the Ottoman failure to develop a competitive maritime power was not due to structural deficiencies but to lack of incentive. This accident of history, as it were, coupled with the European discovery of the resource-full New World, put the Ottomans, and the Muslim world more generally, at an economic disadvantage vis-à-vis their European competitors and created a gap that, with the passage of time, proved difficult to bridge. I am not trying here to give a definitive explanation of the causes of decline but simply suggesting that the decline of the rational sciences in the Muslim world, along with other intellectual activities, was a symptom of complex and historically specific social, political, and economic factors and not the result of an inevitable unfolding of culture.

Without positing a linear causal relationship, let me note that the relative decline in scientific activity, whether reversible or not, is often coupled with a return of interest in cosmology; only this time the traditional philosophical cosmology was replaced with a religious/Sufi one. Thus, for example, in the period of plague and protracted wars described by Ibn Khaldūn, Ibn Bannā' al-Marākishī (d. c. 1321), the most famous North African mathematician and scientist of the fourteenth century, includes Sufi-inspired cosmological interpretations in his otherwise advanced mathematical works.[9] In contrast to the systematic Aristotelian cosmology, the new mystical cosmologies were random and undisciplined exercises of imagination: they were "unscientific," idiosyncratic literary exercises that teach us nothing about the celestial region and the rules that govern the cosmos and its motions.[10] Yet here, too, the introduction of religious and mystical cosmologies into scientific works was a symptom of the decline of the sciences and not their cause. Moreover, decline, when it happened, influenced all forms of intellectual production and was not caused by the rise of one form (religious sciences) at the expense of another (rational sciences).

Uneven and reversible decline patterns are traceable into the early nineteenth century. As Ekmeleddin Ihsanoglu has shown, science activity in the Ottoman empire continued into the eighteenth century, when the earlier traditions of science gradually shifted in favor of attempts to adopt European sciences. The gradual shift in focus was paralleled by state-strengthening modernization efforts that included the army, along with adjunct reforms in the administrative and fiscal systems, law, and education.[11] Many of the eighteenth-century modernization efforts were in response to European threats and often involved emulating the Europeans in order to compete with them. On almost all fronts, however, indigenous attempts to modernize had roots dating as far back as the fourteenth century. The Ottomans already used gunpowder weapons on a large scale then, and imposed new forms of military organization. The efforts to modernize the army received a major impetus in the New Order (*Niẓām Jadīd*) introduced under Sultan Maḥmūd II (r. 1808–39). Under the sultan, the Ottomans relied heavily on European (especially French and German) military advisors and technologies, yet their dependency, which became endemic later, was still driven in the early parts of the nineteenth century by internal needs.

Another notable promoter of modernization was Muḥammad 'Alī Pāshā (r. 1805–48), the ambitious Ottoman governor who established an autonomous polity in Egypt and eventually tried to seize power from the Ottomans. Muḥammad 'Alī's military threats to the Ottomans led to the first alliance of its kind between the Ottoman sultans and the Europeans (mainly British); together, Ottoman and British armies defeated Muḥammad 'Alī's army in Syria and checked its northward advance toward Anatolia. In his attempt to strengthen his state, Muḥammad 'Alī focused on the transfer of technology and technological knowledge from Europe to Egypt. He built military and textile factories, sent more than four hundred students to study science and military tactics in European universities, established technical engineering schools in Egypt, and launched a systematic effort to translate and publish scientific books. In the end, his experiment was aborted. When he was defeated in Syria, European powers recognized the autonomy of his and his descendants' rule in Egypt,[12] but part of the deal allowing him to retain control over Egypt was to dismantle the factories he had built and ship them back to Europe.[13] What the long-term impact of Muḥammad 'Alī's changes might have been if they had lasted is not clear. What is clear, however, is that the decline of science and technology was closely related to political, economic, and cultural dependency, created and buttressed in the course of the encounter with colonialism. What might have been a temporary decline in scientific and technological production in the Muslim world was compounded by the rapid growth of European science, a science nurtured by an emerging capitalism and an expanding, powerful technology. Falling behind in the sciences made it hard for Muslim countries to adopt and develop new technologies, and minimized the chances for further developments that depended on the new technologies. Together, these factors created a gap between Islamic science and European science that colonialism further sustained and enlarged.

Even when driven by competition with Europe, the early modernization projects were apparently free of feelings of cultural inferiority; they seem to have proceeded on the assumption that a balance of power between the Muslim world and its European competitors could be restored. Later, in the 1880s, following major incursions by European colonial powers—for example, the French occupation of Algeria in 1830 and Tunisia in 1881 and the formal British occupation of Egypt in 1882—many Muslim intellectuals expressed in their writings an awareness of European superiority and of the need for an Islamic renaissance to pull the Muslim world out of its state of backwardness. At this time, Europe's industrial age was more than a century old, and a series of fresh conquests resulting from new scientific and technological discoveries and inventions was giving European powers fresh impetus, as was the massive growth of European capitalism and the resulting global colonial project, which spread over much of Asia and Africa and included most of the Muslim world. As a result of this new wave of European expansion, Muslims were exposed to, and became aware of the products of, the industries and liberal ideologies of European powers. In the late nineteenth century, a question asked by many Muslim intellectuals was, Why did the Muslims fall behind while Europe progressed? In the early twentieth century, this question was often reformulated: Why did Muslims fail to achieve a renaissance despite their awareness of a need for one? What structural factors thwarted the Islamic modernization efforts? Invariably, science was invoked in discussions of decline and ways of overcoming it. And despite the bleak conditions for the practice or progress of science in the contemporary Islamic world, most Muslim intellectuals, and people with a broad range of opinions generally, proclaim its neutrality and value. Yet the outcome suggests that these discussions have had little impact on the development of a contemporary scientific culture in Muslim societies.

The Status of Science in Modern Muslim Societies

A historical rather than essentialist view of decline suggests, among other things, that the partial decline of scientific activity in fourteenth-century North Africa was different from the slowdown in scientific activity in, say, the seventeenth century, relative to accelerated scientific activities in Europe, and both declines were fundamentally different from the seemingly entrenched decline in the colonial and postcolonial periods. As I have already suggested, a main difference between the premodern and the modern cases is that, generally speaking, the modern Islamic discourses on science do not inform and are not informed by a living, productive Islamic culture of science (or technology). The science that is the subject of modern Islamic commentaries is produced outside the Muslim world.

Scholars who monitor the status of science and technology in the modern Muslim world often note the lack of accurate statistical information and the need to enhance science and technology indicators (STI) monitoring.[14] Irrespective of the accuracy of the data, all studies concur in their assessment of the dismal state of scientific and technological production in the Muslim world.[15] Despite the richness of many Muslim countries in primary resources, these countries continue to rely, almost exclusively, on imported technologies.[16] As consumers and not producers of science and technology, Muslim societies have shaped their science education policies in light of this fundamental fact of dependency. As far back as the nineteenth century, generations of scientists have been trained in the technical and scientific skills needed to consume imported technologies but not to reproduce them. Most qualified scientists are foreign trained, and specialize in topics that are not necessarily relevant to local needs. On the other hand, despite recent dramatic increases in enrollment in local universities that provide higher education, these universities usually produce nonfunctional scientists who cannot compete with foreign-trained scientists and who often cannot even find employment in their fields of expertise.[17] And because science and technology policies are often imposed from above instead of being socially driven, they are largely subsumed under and restricted by economic goals and do not translate into autonomous science policies. Most technological projects are executed by foreign contractors, and in the execution of the contracts, little effort is made to transfer technological know-how to local scientists.[18]

One consequence of these combined factors is that, whatever indicators are used to measure it, the science research base in contemporary Muslim societies is extremely impoverished.[19] The section on science in the UNDP Arab Human Development Report of 2002 notes that Arab countries "have some of the lowest levels of research funding in the world. . . . R&D expenditure as a percentage of GDP was a mere 0.4 for the Arab world in 1996, compared to 1.26 in 1995 for Cuba" and compared to a global average of over 2.36 percent.[20] This low investment in research and development keeps the scientific community small, marginalized, and unproductive. In terms of the number of scientific publications, the average output of the Arab world per million inhabitants is roughly 2 percent of the output of an industrialized country and less than 1 percent of the output of the United States.[21]

Following a recommendation by the First Islamic Conference of the Ministers of Higher Education and Scientific Research, held in Riyadh in 2000, ISESCO, the Islamic counterpart of UNESCO, established a special Center for Promotion of Scientific Research (ICPSR). All parties involved in establishing the center explicitly recognized the dismal state of science and technology in Muslim countries. The new center was charged with promoting advanced research in science and technology (S&T), providing a forum for the exchange of scientific knowledge, initiating and supporting individual research projects in priority areas, and so on. The identified priority areas include biotechnology, agriculture, information technology, geographic information systems and remote sensing, water management, medicine and pharmaceuticals, renewable energy, and environmental protection and sustainable development. So far, the center has offered a few research grants and published four issues of a journal devoted to scientific and technological research, but it is still too early to judge the center's effectiveness or the quality and impact of the projects it has sponsored.[22] Whether it is successful or not, we are still looking at a centralized, top-bottom initiative that is subsumed under the larger economic development policies of member states. Moreover, the identified priorities seem to be taken out of a textbook; there is no evidence that they are informed by local needs and capacities.

On many levels, then, the engagement with science in the modern Muslim world, from the 1880s and to the beginning of the twenty-first century, has been an engagement with an outside phenomenon, which has had no effect on the actual production of science in Muslim societies. It is therefore reasonable to say, at the outset, that in all varieties, the modern Islamic attitudes toward science have been, in effect, responses to Western science and not epistemological reflections on an indigenous aspect of Islamic culture. The reality of the absence of a modern scientific culture has shaped much of the Islamic discourse on modern science, to which we now turn.

Scientific Positivism and the Technologies of Power

In 1883, Ernest Renan, the leading French Orientalist of his time, published an article (first delivered as a lecture at the Sorbonne) entitled "L'islamisme et la science."[23] In this and other essays, Renan argued that Arabs (and Semites more generally) are by nature hostile to philosophy and science and that Islam is also hostile to science and is, in his words, "the heaviest chain that humanity has ever borne."[24] The same year, the leading Muslim thinker and pan-Islamist activist of the time, Jamāl al-Dīn al-Afghānī (d. 1897), published a response to Renan in which he concedes that "all religions are intolerant, each in its way," but adds that Europeans had a head start on Islam; and just as European rational thinking was able to overcome Christian dogma, Muslims, too, given the chance, would eventually be able to overcome the dogma of their religion and revive scientific thinking. There is so much I could say about Al-Afghānī and how his astonishing response relates to his other ideas and agendas. For our purposes, what is most striking is that the most important Muslim figure of the time adopts, without questioning, the post-Enlightenment, positivistic outlook that categorically opposes science and religion. Al-Afghānī published several other essays in which he promoted educational reform and argued for the benefits of philosophy and science. More important, however, Al-Afghānī published essays criticizing the Muslim materialists and, in particular, what he called the Neichiri sect: the naturalist followers of the Indian thinker Sir Sayyid Aḥmad Khān (d. 1898).[25]

I will come back to Al-Afghānī's critique of Khān, but first let me underscore the positivist understanding of science shared by most Muslim thinkers in the late nineteenth and early twentieth centuries. In fact, this unquestioning understanding of science as a positive value was prevalent throughout the period and continued unabated into the modern period. Unlike Al-Afghānī, later Islamists did not blame Islam for the stagnation of science, but they argued that a scientific and technological revival was required to revive Muslim societies. In an essay-type message sent to the Egyptian king Fārūq and his prime minister, Muṣṭafā al-Naḥḥās Pāshā, in 1947, Ḥasan al-Bannā (d. 1949), the founder of the Egyptian Muslim Brotherhood, says: "Just as nations need power, so do they need the science, which will buttress this power and direct it in the best possible manner. . . . Islam does not reject science; indeed, it makes it as obligatory as the acquisition of power, and gives it its support. . . . The Qur'ān does not distinguish between secular and religious science, but advocates both," and "God commands mankind to study nature and He prompts them to it."[26] Similar assessments of the need for science are echoed in the works of Sayyid Quṭb, the most influential twentieth-century Islamist and the leader of the Egyptian Muslim Brotherhood; in the works of Abū al-A'lā al-Mawdūdī, the founder of Jamā't-i-Islāmī, the main Islamic party in Pakistan; and in the works of many other Islamists.[27] Despite their critiques of Western imperialism and of Western culture and morality, these Islamists accepted Western science on the grounds that science is a historical product whose development and ownership are universal, not cultural. They assume, unreflectively, the monolithic nature of science and the scientific method.[28] More important, science and technology were not distinguished in their discourse; both were seen as requisite instruments of power, ones that Muslims needed in their political and social struggles.[29]

This positivistic understanding of modern science as an instrument of power rather than a system of thought with its own epistemological assumptions characterizes even the critical discussions of modern science in the late nineteenth century. The two areas in which an Islamic stand on modern science began to be articulated were the new Copernican astronomy and Darwinism.

The New Astronomy

The first discussions of Copernican astronomy were held in the Arab world during the second half of the nineteenth century. Almost all such discussions focused on the question of the mobility of the earth, and none addressed the mathematical aspects of this astronomy.[30] Without exception, none of these discussions included reference to the Islamic tradition of mathematical astronomy that had informed Copernicus, suggesting an almost total lack of knowledge and appreciation of what has become a mostly forgotten Islamic scientific tradition.[31] Not even the controversies over the new astronomy were homegrown; they seem to have been mechanical transpositions of earlier Western controversies brought to the Muslim world in missionary schools.

The Syrian Protestant College (which later became the American University of Beirut) was founded in 1866, and the competing French Jesuit Saint Joseph University, also in Beirut, was founded in 1875. In 1876, the Syrian Protestant College started to issue an influential journal called *Al-Muqṭaṭaf* (it later moved to Egypt) whose declared mission was to promote modern science and technology. The journal was not a scientific journal in the technical sense of the word, but it advocated scientific values. Soon, the Jesuits started to publish a similar journal, entitled *Al-Bashīr.* A little earlier, a Maronite Christian intellectual named Salīm al-Bustānī also started to publish a journal—his was entitled *Al-Jinān*. All three journals included extensive discussions of science and attempted to identify and promote what they considered to be the values of modern science. Between 1872 and 1876, *Al-Jinān* published a series of articles describing, in very general terms, the new astronomy, and in 1876 it published an attack on this astronomy by a Palestinian Christian, Naṣīr al-Khūrī. *Al-Muqṭaṭaf* published what amounted to a response to this attack the same year. Its author was one of the founders of the journal, Ya'qūb Ṣarrūf. Ṣarrūf's defense was superficial and exhibited little scientific or mathematical understanding of the new astronomy. A few months later, *Al-Muqṭaṭaf* published a response to Ṣarrūf's article by the Greek Orthodox Patriarch of Antioch, Archimandrite Gabriel Jibāra. Jibāra provided a doctrinal refutation of the new astronomy based on verses of the Old Testament; the miracle of Joshua ordering the sun to stand still indicated, he argued, that the sun moves. As we can see, the whole debate was initiated by Christian intellectuals and echoed earlier European debates and controversies.[32]

Following Jibāra's article, however, 'Abd Allāh Fikrī, the Muslim Egyptian deputy of education, joined the debate. He wrote a letter to *Al-Muqṭaṭaf* asserting that the concept of the mobility of the earth was not a novel idea and that one can find it in earlier Islamic writings. As evidence, Fikrī referred to Al-Ghazālī, who maintained that the study of astronomy was not against Islamic law and that the conclusions of astronomers should not be construed as being in opposition to religion as long as they recognized that the world is created by God. Jibāra wrote back to the editors of *Al-Muqṭaṭaf* and accused them of soliciting a Muslim's opinion when they failed to find an approving Christian one. The whole discussion ended with an article published by the editors of *Al-Muqṭaṭaf* reminding Jibāra that his position regarding the opposition between the new astronomy and the Holy Scripture was not the position of his own church and suggesting that his superiors would not think he was serving the interests of their congregation. Jibāra wrote no more.[33] Six years later, a Syrian Muslim from Ḥama published in *Al-Muqṭaṭaf* another defense of the new astronomy; he recapitulated Fikrī's arguments, together with those of 'Aḍuḍ al-Dīn al-Ījī (d. after 756); and for the first time mentioned the name of Copernicus in association with the new astronomy.

The Christian writers in the missionary schools who apparently initiated the debate over the Copernican astronomy stood apart from the general Christian population, who tended to be more traditional than the graduates of the schools. The few Muslims who spoke up sided with the progressive element in the Christian community and were quick to work on the problem of harmonizing the new astronomy with religion. But the debate was limited to the question of the earth's motion and did not deal with the mathematical aspects of the new astronomy, nor was the Islamic astronomical tradition of the thirteenth to sixteenth centuries invoked, suggesting that this tradition was forgotten and could not be easily recalled. But if Muslim intellectual circles paid little attention to the new astronomy, the case was quite different with Darwinism.[34]

Darwinism

Missionary schools also initiated the discussions of Darwinism, but those discussions differed in two main respects from the discussions of the new astronomy. For one thing, the Jesuits operating in Lebanon who were indifferent to the first debate participated in the second and tended to oppose Darwinism, as did their journal *Al-Bashīr,* whereas the graduates of the Protestant school, expressing their views in *Al-Muqṭaṭaf,* tended to side with Darwin against the administrative body of their university. For another thing, Darwinism invoked a much wider response from Muslims, both in support and in opposition, than did Copernican astronomy. Two major contributors to the debate were Shiblī Shumayyil (1850–1917), a young Christian physician who introduced Darwin to the Arab east, and Al-Afghānī, who was one of the earliest and most influential Muslim opponents of Darwinism. Shumayyil was introduced to Darwinism while still a student at the medical school of the Syrian Protestant College. In 1882 the College Board of Trustees forced a certain Professor Edmund Lewis to resign after approvingly referring to the theory of evolution in his commencement address. The scandal left its marks on Shumayyil, then a student. In the 1880s he published a series of articles in *Al-Muqṭaṭaf* (which had then moved to Egypt), and in 1910 the whole series was published in a book entitled *The Philosophy of Evolution and Progress* (*Falsafat al-Nushū' wal-Irtiqā'*).[35]

Although Shumayyil was a scientist by training, his interest in Darwinism was primarily social and philosophical. "My purpose in this treatise," he writes, "is to establish the materialist philosophy on solid scientific grounds"—that is, to provide proofs of the materiality of the world that could not have been created out of nothing. Shumayyil argues that "the natural sciences are the source of all sciences, and they must be so, and they must precede all other sciences." Shumayyil attacks all traditional disciplines, including philosophy, theology, linguistic sciences, literature, and law as worthless and maintains that progress is possible only through the natural sciences. He also proposes abolishing law schools and destroying books on economics and philosophy.[36] Christian authors published several criticisms of Darwinism, but its most influential Muslim critic was Al-Afghānī.[37]

The essay that contains Al-Afghānī's attack, which is directed specifically against the Indian advocate of evolution, Sir Sayyid Aḥmad Khān, is entitled "The Truth about the Neichiri Sect and an Explanation of the Neichiris." Originally published in Persian in 1881, it was later translated into Arabic as *Al-Radd 'alā al-Dahriyyīn* (A Refutation of the Materialists). Khān, who was educated in England, founded the Neichiri ("from nature") sect in India, and it was the political implications of the establishment of the sect that were Al-Afghānī's main concern, not the scientific aspects of Darwinism. Advocates of Neichirism were champions of close cooperation with colonial Britain, and Khān and the Neichiris of India advocated complete westernization at a time when anticolonial sentiment was increasing in the Muslim world.[38]

One of Al-Afghānī's main objections to the Neichiris was

their introduction of yet another Islamic sect, representing a further division of the Muslims who needed, more than ever before in their history, to be united. Al-Afghānī also argues that the laws of nature do not supply a moral code, which Muslim societies need to cohere and revive. By supporting the idea of evolution, the Neichiris were undermining the possibility of an Islamic social resurrection. Al-Afghānī even accuses Khān of being a British agent who was "misdirecting the minds of Muslims, sowing disunion among them, and instigating enmity between the Muslims of India and other Muslims, especially the Ottomans."[39] Unlike European naturalists, Al-Afghānī adds, Khān and his followers couple their naturalist beliefs with disregard for the national interests of their homelands and a readiness to submit to foreigners, all of them thereby serving as agents of British colonialism.[40] His political attacks aside, all of Al-Afghānī's arguments against evolution were simple and reflected the widespread misunderstanding of Darwinism. Muḥammad ʿAbdu, Al-Afghānī's associate and student, later noted that Al-Afghānī wrote the refutation when he was passionately angry with the advocates of complete westernization.[41] A few years later, Al-Afghānī went back to the subject to discuss it less passionately and more systematically. In *Khāṭirāt Jamāl al-Dīn al-Afghānī* (The Memoirs of Jamāl al-Dīn al-Afghānī), Al-Afghānī seems to accept the notion of natural selection as it applies to plants and animals and also ideas.[42] He contends that this principle was known and implemented in the pre-Islamic and Islamic cultures long before Darwin. But Al-Afghānī retained his original rejection of human evolution.[43]

Despite his unqualified defense of philosophy and science, in this instance Al-Afghānī conceived of Darwinism as an instrument of cultural imperialism. In his attack on the Neichiri sect, Al-Afghānī was less concerned with Darwin's theories than with Khān's naturalistic interpretation of the Qur'ān. Khān argues that God created a book, the Qur'ān, but also created nature; and while the Qur'ān admits allegorical interpretations, the laws inscribed in God's book of nature are certain and final. As such, Khān argues, whenever natural laws and the statements of the Qur'ān appear to contradict each other, the Qur'ānic statements must be interpreted allegorically. Khān also suggests that traditional Islamic beliefs are responsible for the backwardness of Muslim societies and that a scientific reinterpretation of these beliefs in light of the laws of nature is required to solve the social problems of Muslim societies.[44]

Clearly, Al-Afghānī's response was primarily political and was directed more at colonialism than at scientific Darwinism. Aside from this politically motivated criticism, however, the early Muslim attitudes to Darwinism were largely ones of indifference or even approval.[45] The early reception of Darwinism may have been somewhat favorable, but a strong creationist movement has appeared in the contemporary Muslim world. Ironically, the Islamic creationist movement originated two decades ago in Turkey, one of the most scientifically advanced Muslim countries.[46] After modest beginnings, in the past decade the movement has gained great momentum and following throughout the Muslim world. In February 2007, schools in France received free copies of a work by the leader of this movement, the leading Muslim advocate of creationism and intelligent design, Adnan Oktar, who writes under the name Harun Yahya.[47] The book sent in this mass mailing is entitled *Atlas of Creation*.[48] Like many of the numerous books and articles attributed to Yahya, it may have been composed by several members of the Science Research Foundation (Bilim Arastirma Vakfi, or BAV), established in 1991 by Oktar/Yahya.[49] Also, like many of the group's other publications, it relies heavily on translations of works by

American Christian fundamentalists.[50] The mass mailing of this book took the French public schools by surprise and alarmed the French Ministry of Education, which scrambled to come up with an adequate response to this unanticipated assault.[51] As with the nineteenth-century debates on Darwinism in the Islamic world, the contemporary rise of a popular Islamic creationist movement suggests, once again, that the Muslim world cannot remain immune to Western debates. Unlike the earlier discussions of science and evolution, however, today's Islamic creationist movement seems to aspire to serve as the international arm of the American creationist movement. Yet, despite its ambitions and popularity, there is nothing specifically Islamic about the derivative arguments of Islamic creationism.[52]

The Qur'ān and Science

In other cases, the Islamic encounter with modern science engendered responses of a more pronounced religious character. In contrast to premodern contentions on the autonomy of science, modern and contemporary Islamic discourses on Islam and science abound with assertions of the relationship between the two. By far the most common treatments of this subject maintain that many modern findings of science were predicted, or at least alluded to, in the Qur'ān, and that these predictions constitute evidence of what is referred to as the scientific miracle (*iʿjāz*) of the Qur'ān.[53] The origins of this approach can be traced back to the nineteenth century; thinkers such as Muḥammad ʿAbdu (d. 1905) and Muḥammad Iqbāl (d. 1938) contended that the Qur'ān and science are in harmony but did not dwell on their relationship. Iqbāl passionately argued that the rise of Islam marked the birth of inductive reasoning and experimental methods, but he did not present the Qur'ān as a source of scientific knowledge, nor did he suggest that someone can arrive at scientific facts through the Qur'ān. Other scholars carried the argument further. The Turkish scholar Bediuzzeman Said Nursi (1877–1960) proposed simple but extremely influential scientific interpretations of the Qur'ān, arguing, for example, that the Qur'ān predicted such things as aviation and the discovery of electricity.[54] Nursi had a significant influence on a large number of Turkish followers and students, including many of the Turkish creationists.[55] Another book that marks a turn in the same direction is by the Egyptian scholar Ṭantāwī Jawharī (d. 1940). In his twenty-six-volume work of exegesis entitled *Al-Jawāhir fī Tafsīr al-Qur'ān al-Karīm* (The Gems in the Interpretation of the Noble Qur'ān), Jawharī made a point that is frequently repeated in the contemporary discourse on the Qur'ān and science—namely, that the Qur'ān contains 750 verses pertaining directly and clearly to the physical universe, whereas it has no more than 150 verses pertaining to legal matters. Jawharī thus calls on Muslims to reverse the order of interest and give priority to the scientific verses, especially since they are now living in the age of science.[56]

In effect, these and other modern approaches to the Qur'ān advocate mixing religious and scientific knowledge. When Muslims were the main producers of science in the world, they did not advocate wedding science and religion. Now, ironically, when Muslim participation in the production of the universal culture of science is dwindling, they call for bringing the two together. As I argued in the previous chapter, classical commentators on the Qur'ān never even hinted that the miracle of the Qur'ān was in its prediction of scientific discoveries, made centuries after the coming of revelation. Nor did the early commentators advocate an understanding of the Qur'ān as a source of scientific knowledge. Both claims abound in contemporary Islamic discourse.

The presumed relationship between science and the Qur'ān

is construed in a variety of ways, the most common of which are the efforts to prove the divine nature of the Qur'ān through modern science. These efforts cover a wide range of activities, among them establishing institutions, holding conferences, writing books and articles, and using the Internet to promote the idea of the scientific miracles of the Qur'ān. Nearly a decade ago, an Internet search found nearly two million postings on Islam and science, most by writers claiming that the Qur'ān's prediction of many of the theories and truths of modern science is evidence of its miraculous nature and its divine origins.[57] Such contentions are not just the product of popular opinion; they are reflected in the writings of many contemporary Muslim intellectuals. In the 1980s the Muslim World League at Mecca even formed a Committee on the Scientific Miracles of the Qur'ān and Sunna (traditions of the Prophet). The committee has since convened numerous international conferences and sponsored various intellectual activities, all aimed at exploring and corroborating the connections between science and the Qur'ān. At a meeting in Cairo in 2003, reported in the mass media, this committee urged Muslims to employ the "scientific truths that were confirmed in the verses of the Qur'ān and that, only recently, modern science has been able to discover" as a corrective to the current misunderstanding of Islam. The scientific miracles of the Qur'ān are, in other words, the only weapon with which contemporary Muslims can defend the Qur'ān, and the Qur'ān offers the only convincing proof of its own truth in this age of science and materiality.[58]

Once a correlation between the Qur'ān and science is asserted, a small extension of the same logic leads to the arbitrary exercise of collecting extra-Qur'ānic facts and discoveries and mining the Qur'ān for statements that seem to correspond to them. That these new scientific discoveries have nothing to do with the Qur'ān never hinders modern commentators, who proudly present these theories as evidence of the Qur'ānic miracle. The Qur'ānic text is read with these so-called scientific facts in mind, without any recognition that this reading is an interpretation of the text conditioned by the assumptions of the interpreters and what they are looking for. Examples of this kind of reading include the assertion that the Qur'ān (21:31) predicts the modern discovery of the role that mountains play as stabilizers (*rawāsī*), and that the Qur'ān (21:30, *al-samāwāt wal-ard kānatā ratqan fafatqnāhumā*, "the heavens and earth were an integrated mass then We split them") contains a condensed version of the big bang theory![59]

The early attempts to interpret the Qur'ān and verify it in light of the discoveries of modern science received added impetus in the last decades of the twentieth century, when attempts were made to articulate the theoretical foundations of a new mode of exegesis that aims not just at providing a scientific interpretation of the Qur'ān but also at illustrating its scientific miracles. The emergence of this scientific tafsīr went through several stages, from listing the verses that admit of a scientific interpretation, to theorizing by laying out the rules for this new mode of exegesis, and to maintaining that i'jāz, the scientific miraculousness of the Qur'ān, is a manifestations of a universal cosmic truth predetermined in the Qur'ān and, further, that scientists can find leads in the Qur'ān that will facilitate future scientific research, presumably by identifying research projects or by finding answers to pending scientific questions.[60] In extreme cases, this approach borders on the cultic, as in the widely circulated genre known as the *i'jāz raqamī* or *'adadī* (numerical i'jāz) of the Qur'ān. This form of numerology asserts a numerical order to the occurrence of certain terms in the Qur'ān, which is seen as yet another miracle.[61]

The scientific exegesis of the Qur'ān invokes the testimony

of science, however randomly constructed, in order to corroborate religious belief. This no doubt reflects a sense of insecurity and a need to vindicate religion in the age of science. A very different modern response to science is tacitly driven by the sense that the discourse about science, and various narratives about the history of modern science and its unique and exclusive rootedness in Europe, is itself often a reflection of a modern civilizational will to dominate. In contrast to the positivist position implied in the scientific exegetical approach, this response provides Islamic criticisms of scientific positivism.

The two main representatives of this critical Islamic approach to modern science are Ziauddin Sardar and Seyyid Hossein Nasr, both of whom have had significant influence among Muslim students in Western academic circles. Sardar argues that "modern science is distinctively Western. All over the globe all significant science is Western in style and method, whatever the pigmentation or language of the scientist. . . . Western science is only a science and not the science. . . . It is an embodiment of Western ethos and has its formation in Western intellectual culture."[62] Elsewhere, Sardar maintains that "Islamic science is based on a set of entirely different assumptions about the relationship between man and man, man and nature, the universe, time and space. Because the axioms of Islamic science are different from those of Western science, and its methods of knowing are more open and all-encompassing, it is a science with its own identity and character." Yet when Sardar ambitiously proposes an "operational model of Islamic science" that "relates science and technology to a set of basic Islamic values" and "embraces the nature of scientific inquiry in its totality," he comes up with the following ten values: *tawheed* (unity), *khilafah* (trusteeship), *ibadah* (worship), *ilm* (knowledge), *halal* (praiseworthy) and *'haram* (blameworthy), *adl* (justice) and *zulm* (tyranny), and *istislah* (public interest) and *dhiya* (waste). This model, Sardar argues, "replaces the linear, enlightenment thinking which is the basis of modern science with a system of knowing that is based on accountability and social responsibility."[63] Sardar's critique underscores the cultural specificity of all forms of knowledge. This critique of science, in its manifold expressions, has been very influential among philosophers of science, but there is nothing specifically Islamic about it, the desire to propose an Islamic epistemology notwithstanding.[64] The ten values of Sardar's model are completely arbitrary; they derive from his own personal and unhinged understanding of Islam, and there are no references whatsoever in classical Islamic scientific or religious sources that relate these religious values to the practice of science.[65]

The other main Islamic critique of scientific positivism, best represented in the writings of S. H. Nasr, takes a different tack. Its advocates question the fundamental metaphysical framework within which modern science operates and attempt to articulate an alternative Islamic framework by positing a dichotomy between ancient and modern sciences and contending that the ancient sciences shared conceptions of sacredness and unity of knowledge.[66] Yet if the distinctiveness of the ancient metaphysical framework lies in the sacredness and unity of knowledge, then it is not clear how Islamic science would be different from, for example, pagan Hellenistic science. Furthermore, as in the epistemological approach, the content of the Islamic metaphysical framework remains unclear. "A truly Islamic science," Nasr contends, "cannot but derive ultimately from the intellect which is Divine and not human reason. . . . The seat of intellect is the heart rather than the head, and reason is no more than its reflection upon the mental plane."[67] This flight into the world of imagination may prove difficult to quantify. More important, however, is that it exactly reverses the main thrust of science in the classical period of Islam, in which reason is treated as a faculty with no separate existence outside human physical existence.

To be sure, the approaches by both Sardar and Nasr are serious intellectual exercises. Even when citing verses of the Qur'ān, however, their arguments remain largely extra-Qur'ānic. Neither approach systematically engages the Qur'ānic text as a whole, or the cultural legacy that endowed the text with its specific historical meanings. Their critical attempts can be viewed as reactions to scientific positivism, but they can also be seen as attempts to restore the sovereignty of metaphysics over science (at least at the discursive level, since these attempts find little, if any, translation at the practical level). This time around, however, the proposed metaphysics is religious, and unlike philosophical metaphysics, it is random, undisciplined, and not capable of producing science.

In the absence of a living scientific culture in the modern Muslim world, we can discuss discourses on science, not science itself. Generated, as it were, in the shadow of modernity, all of the discourses considered here are reflections on science, modernity's most salient ingredient. While some approaches to science reflect a desire to appropriate this instrument of power, others reflect a desire to extract scientific modernity from westernization. Yet the newly constructed Islamic discourse on science, whatever approach is taken, is not rooted in a historical understanding of the relationship between Islam and science. On one level, this is understandable. However defined, modern science has engendered, and continues to engender, multiple and intense responses among Muslims and non-Muslims alike. As an instrument of power, science is a tool that perpetuates the subjugation and dependency of Muslims, but it also embodies the hope for their future recovery. The challenges posed by the modern culture of science have no parallel in premodern societies. In thinking about modern science Muslims today have to confront challenges that were not addressed in the classical period of Islam. Yet the desire to articulate contemporary critical concerns about science in Islamic language cannot conceal the radical departure of these modern articulations from the classical ones. A shared trait of all modern Islamic discourses on science is their presenters' ignorance of, indifference to, or even outright abuse of history. To be sure, the normative past need not be indefinitely preserved or reproduced. When confronted with the realities of modernity, including not just highly complex technologies but also developed and complex ethical debates about science and technology, Muslims cannot formulate their views on science in isolation from the world around them, nor is it desirable for them to do so. But a departure from a normative past ought to be informed and deliberate if it is to have any meaningful and lasting impact on the building of a desired future.

Notes

Chapter 1. Beginnings and Beyond

1. The Ka'ba is the cubical structure in Mecca at the center of the Islamic pilgrimage ritual.

2. More precisely, the zenith is the imaginary point of the heavens directly above an observer on the earth through which pass the prime vertical and meridian circles.

3. In other words, the azimuth is an arc of the horizon between the meridian of the observer—that is, the imaginary great circle passing through the zenith of the observer's locality and the north and south poles—and the vertical circle passing through the zenith of Mecca.

4. David King, "Science in the Service of Religion: The Case of Islam," in King, *Astronomy in the Service of Islam* (Aldershot, UK: Variorum, 1993), 261.

5. *Qiblat ahl Fās* means, more precisely, the "qibla of the people of Fes."

6. In the following pages I do not aim to provide an exhaustive treatment of the debate on the direction of the qibla of Fes or on the ways through which this debate was resolved. A separate monograph would be needed to account for the many subtle and context-specific aspects of the debate.

7. My overview is based on the following texts: Abū 'Alī Ṣāliḥ al-Maṣmūdī (14th c.), *Kitāb al-Qibla*, published under the title *La Alquibla en al-Andalus y al-Maghrib al-Aqsa*, ed. and trans. Monica Rius (Barcelona: Institut "Millás Vallicrosa" d'Història de la Ciència Àrab, 2000); Al-Maṣmūdī, *Risāla fī*

Ta'yīn Jihat al-Qibla, ms. in Al-Khizāna al-Ḥasaniyya (Royal Library), Rabat, Majmū' 12399z; Al-Maṣmūdī, *Risāla fī Ittijāh al-Qibla fī Ba'd al-Buldān*, ms. in Al-Khizāna al-Ḥasaniyya, Majmū' 6999, fols. 23–25. Al-Maṣmūdī provides extensive quotations from earlier scholars on the subject, including quotations from Abū 'Alī al-Mittījī (12th c.), *Kitāb Dalā'il al-Qibla;* 'Abd al-Raḥmān al-Tājūrī al-Maghribī al-Ṭarābulsī (d. 1590), *Tanbīh al-Ghāfilīn 'an Qiblat al-Ṣaḥaba wal-Tabi'īn*, Al-Khizāna al-Ḥasaniyya, ms. 10153; Al-'Arabī b. 'Abd al-Salām al-Fāsī (end of 17th c.), *Shifā' al-Ghalīl fī Bayān Qiblat Ṣaḥib al-Tanzīl*, Al-Khizāna al-Ḥasaniyya, ms. 6588. The last treatise was written in response to a critique of Al-Tājūrī by 'Abd al-Raḥmān b. 'Abd al-Qādir al-Fāsī (d. 1685). 'Abd al-Raḥmān's son Muḥammad (d. 1722) also wrote a critique of Al-Tājūrī, *Iqāmat al-Ḥujja wa Iẓhār al-Burhān 'alā Ṣiḥḥat Qiblat Fās wa mā Wālāha fī al-Buldān*. The above texts, all written between the twelfth and the early eighteenth centuries, provide detailed accounts of opinions on both sides of the debate, including views of astronomers as well as religious scholars whose legal rulings (*fatwas*) were solicited on this matter.

8. Al-Mittījī, quoted in Al-Maṣmūdī, *Kitāb al-Qibla*, 15.

9. Al-Maṣmūdī, *Kitāb al-Qibla*, 4–5.

10. Al-Maṣmūdī, *Kitāb al-Qibla*, 12.

11. Al-Maṣmūdī, *Kitāb al-Qibla*, 13.

12. In *Risāla fī Ittijāh al-Qibla*, Al-Maṣmūdī maintains that calling for people in Morocco to pray while facing south is tantamount to calling on people to commit a sin.

13. Al-Maṣmūdī, *Kitāb al-Qibla*, 16–19.

14. Al-Tājūrī is mentioned in 'Abd al-Wahhab al-Sha'rani's biographical dictionary of Sufis, *Lawāqiḥ al-Anwār fī Ṭabaqāt al-Akhyār al-Mashhūr bi al-Ṭabaqāt al-Kubrā* (Cairo: Maktabat al-Thaqāfah al-Dīnīya, 2005).

15. Fes is one of the most important cities in Morocco, and for much of the history of Islam in North Africa it was the political capital of the country and also its scholarly capital. In terms of Islamic education, Al-Qarawiyyīn is the Moroccan equivalent of the Azhar University in Egypt.

16. Including the Imam Idrīs b. Idrīs b. 'Abd Allāh b. al-Ḥusayn b. 'Alī (d. 828), a member of the family of the Prophet.

17. These, according to Al-Fāsī, include Al-Abharī and Al-Qarāfī. Al-Fāsī, *Shifā' al-Ghalīl*, fol. 6; hereafter cited in the text.

18. Latin, Zarqallu. The Zarqālī plate is a universal astrolabe with only one plate, named after its inventor, Ibrahim al-Zarqālī (Toledo, 11th c.).

19. See, for example, Dimitri Gutas, who maintains in his study of the

translation movement that "it was the development of an Arabic scientific and philosophical tradition that generated the wholesale demand for translations from the Greek (and Syriac and Pahlavi), not, as is commonly assumed, the translation which gave rise to science and philosophy." Dimitri Gutas, *Greek Thought, Arabic Culture: The Graeco-Arabic Translation Movement in Baghdad and Early 'Abbāsid Society (2nd–4th / 8th–10th Centuries)* (London: Routledge, 1998), 137, 150.

20. The question of the influence of new ideas applies not only to the influence of Islamic scientific discoveries on Latin European science but also to the role of early ideas in later developments of particular scientific traditions. For example, Roshdi Rashed provides compelling accounts of several stages in the development of the science of algebra, from the modest but conceptually distinctive beginnings with Al-Khwārizmī's (fl. 830) seminal work, *Al-Jabr wal-Muqābala,* to the works of 'Umar al-Khayyām (1048–1131) and Sharaf al-Dīn al-Ṭūsī (d. 1274). Another example is Al-Kāshī's (d. 1436) *Miftāḥ al-Ḥisāb,* the culminating work of a continuous tradition of arithmetic within the school of Abū Bakr al-Karajī (late 10th c.); it includes the results of many mathematicians within the school, such as the works on decimal fractions by Al-Samaw'al al-Maghribī (12th c.), *Al-Bāhir fī al-Jabr* and *Al-Qawāmī fī al-Ḥisāb al-Hindī.* See Roshdi Rashed, *Tarīkh al-Riyādiyyāt al-'Arabiyya bayna al-Jabr wal-Ḥisāb* (The History of Arabic Mathematics: Between Algebra and Arithmetic) (Beirut: Markaz Dirāsāt al-Waḥda al-'Arabiyya, 1989), xxx, 112.

21. For example, numerous works are dedicated to dismissing the discovery by Ibn al-Nafīs (13th c.) of the pulmonary transit of the blood as an out-of-context, happy guess with no scientific significance. To be sure, the context and meaning of Harvey's discovery of the pulmonary circulation of the blood is different from the context and meaning of Ibn al-Nafīs's discovery, but this does not mean that the latter discovery is out of context. For a corrective to the view that Ibn al-Nafīs's discovery had no context of its own see the excellent dissertation by Nahyan Fancy, who situates it in the context of Ibn al-Nafīs's culture and in relation to his other philosophical and religious ideas. Nahyan Fancy, "Pulmonary Transit and Bodily Resurrection: The Interaction of Medicine, Philosophy and Religion in the Works of Ibn al-Nafis (d. 1288)" (Ph.D. diss., University of Notre Dame, 2006). Similar assertions are made in connection to many other discoveries. It was said, for example, that the pioneering work on optics by Ibn al-Haytham (d. 1039) was not appreciated in the Muslim world and was appreciated only later on in Europe; the same is said of Ibn al-Shāṭir's (d. 1375) new astronomical models and even of the philosophical ideas

of Ibn Rushd (Averroës; d. 1198). Evidence of Ibn al-Shāṭir's influence can be inferred from the inclusion of his works in several pedagogical collections—for example, Al-Khizāna al-Ḥasaniyya, Majmu' 1676—and a reference to a different treatise in ms. 2723, Al-Khizāna al-Ḥasaniyya.

22. E. S. Kennedy, colleagues, and former students, *Studies in the Islamic Exact Sciences*, ed. David King and Mary Hellen Kennedy (Beirut: American University of Beirut, 1983).

23. George Saliba, *A History of Arabic Astronomy: Planetary Theories during the Golden Age of Islam* (New York: New York University Press, 1944); and, more recently, Saliba, *Islamic Science and the Making of the European Renaissance* (Cambridge: MIT Press, 2007). Other collections of specialized studies with useful general overviews include Julio Samso, *Islamic Astronomy and Medieval Spain* (Aldershot, UK: Variorum, 1994). In addition, titles of editions, translations, and studies of several important classics of Arabic astronomy can be found in the bibliography of Roshdi Rashed, edited in collaboration with Régis Morelon, *Encyclopedia of the History of Arabic Science*, vol. 1: *Astronomy: Theoretical and Applied;* vol. 2: *Mathematics and the Physical Sciences;* vol. 3: *Technology, Alchemy, and the Life Sciences* (London: Routledge, 1996); see, in particular, entries by E. S. Kennedy, George Saliba, David King, Régis Morelon, and Jamil Ragep.

24. See, for example, King, *Astronomy in the Service of Islam;* David King, *Islamic Mathematical Astronomy* (London: Variorum, 1986); and King, *Islamic Mathematical Instruments* (London: Variorum, 1987). On astronomical observations, Aydin Sayili, *The Observatory in Islam* (Ankara: Türk Tarih Kurumu, 1960), remains a classic.

25. Rashed has produced several critical editions and translations of and commentaries on Arabic mathematical texts in the disciplines of algebra, geometry, arithmetic, numerical analysis, infinitesimal mathematics, and mathematical optics. An overview of some of Rashed's findings is in Roshdi Rashed, *The Development of Arabic Mathematics: Between Arithmetic and Algebra,* trans. A. F. W. Armstrong, Boston Studies in the Philosophy of Science Series, 156 (Dordrecht: Kluwer, 1994).

26. See M. Ullmann, *Die Medizin im Islam* (Leiden: Brill, 1970). Another book by the same author, M. Ullmann, *Islamic Medicine,* Islamic Surveys, 11 (Edinburgh: Edinburgh University Press, 1978), is a concise survey of Islamic medicine. Another useful overview is the introduction by Michael Dols to 'Ali Ibn Riḍwān, *Medieval Islamic Medicine: Ibn Ridwān's Treatise "On the Prevention of Bodily Ills in Egypt,"* trans. Michael Dols; Arabic text ed.

Adil S. Gamal (Berkeley: University of California Press, 1984). See also Lawrence Conrad, "The Arab-Islamic Medical Tradition," in Conrad et al., eds., *The Western Medical Tradition: 800 BC to AD 1800* (Cambridge: Cambridge University Press, 1994), 93–138. A recent historical overview is Peter Pormann and Emilie Savage-Smith, *Medival Islamic Medicine* (Washington, DC: Georgetown University Press, 2007). A collection of several influential essays is Max Meyerhof, *Studies in Medieval Arabic Medicine: Theory and Practice* (London: Variorum, 1984). On Arabic pharmacology see the works by Ibrahim Ibn Murad, especially his *Buhuth fī Tarikh al-Tibb wal-Saydala 'ind al-'Arab* (Beirut: Dār al-Gharb al-Islāmī, 1991). On hospitals in the Muslim world, the most comprehensive work to date remains Aḥmad 'Īsā, *Tārīkh al-Bīmāristānāt fī al-Islām* (Damascus: Al-Maṭba'a al-Hāshimiyya, 1939).

27. For other sciences see, for example, Ibn al-Haytham, *The Optics of Ibn al-Haytham: Books I–III, on Direct Vision*, 2 vols., trans. A. I. Sabra, Studies of the Warburg Institute, 40 (London: Warburg Institute, University of London, 1989). See also the useful collection of articles in A. I. Sabra, *Optics, Astronomy and Logic: Studies in Arabic Science and Philosophy* (Aldershot, UK: Variorum, 1994). On technology see Ahmad Y. al-Hasan and Donald Hill, *Islamic Technology: An Illustrated History* (Cambridge: Cambridge University Press, 1986).

28. Fuat Sezgin, *Geschichte des Arabischen Schrifttums*, 9 vols. (Leiden: Brill, 1967–); Gerhard Endress, "Die Wissenschaftliche Literatur," in *Grundriss der Arabischen Philologie*, vol. 2, ed., Helmut Gatje; vol. 3, supplement, ed., Wolfdietrich Fischer (Wiesbaden: Reichert, c. 1982), 2:400–506; 3:3–152; Ullmann, *Die Medizin im Islam;* and M. Ullmann, *Die Natur und Geheimwissenschaften im Islam* (Leiden: Brill, 1972).

29. Gutas, *Greek Thought, Arabic Culture;* and George Saliba, *Al-Fikr al-'Ilmī al-'Arabī; Nash'atuhu wa Taṭawwuruhu* (Balamand, Lebanon: Balamand University, 1998).

30. Among other things, both provide a thorough critique of Ignaz Goldziher's standard thesis about the conflict between Islam and science as articulated in I. Goldziher, "The Attitude of Islamic Orthodoxy towards the Ancient Sciences," in Merlin Swartz, ed., *Studies in Islam* (New York: Oxford University Press, 1981), 185–215.

31. See, for example, Saliba, *Al-Fikr al-'Ilmī*, 52–72. Saliba shows that the word *dīwān*, often wrongly translated as "register of salaries," in fact referred to a complex set of bureaucratic and administrative operations that included land surveying, inheritance and calendar computations, irrigation

technologies, weights and measures, geometry, and so on. Saliba argues that "the translation movement was primarily an administrative movement" instigated by professional competition over administrative positions, which became increasingly dependent on the acquisition of higher specialized knowledge. Saliba, *Al-Fikr al-ʿIlmī*, 62–72.

32. To wit, the development of astronomy before the beginning of the translation movement and independent of astrology. See Saliba, *Al-Fikr al-ʿIlmī*, 83, 87.

33. For the notion that translation is a process that transforms the two languages involved, not a linear and unidirectional movement from one language to another, see, for example, Scott L. Montgomery, *Science in Translation: Movements of Knowledge through Cultures and Time* (Chicago: University of Chicago Press, 2000): "Translation means creating an embrace between languages, the producing of an offspring" (291).

34. For example, Gutas, *Greek Thought, Arabic Culture*, 137; and Saliba, *Al-Fikr al-ʿIlmī*, 23–72. Saliba notes that no original production survives from these so-called pockets of knowledge, nor do historical sources mention such works. Both Gutas and Saliba note that in the sixth and seventh centuries the existence of Greek scientific works in Byzantium did not produce a scientific culture, although there was no language barrier preventing the reading of these works.

35. Saliba notes that minimum knowledge on the receiving end was needed to understand scientific sources, recognize their value, and integrate the knowledge into a scientific environment. Saliba, *Al-Fikr al-ʿIlmī*, 25.

36. Saliba argues that the science of astronomy was developed before the beginning of translation, as evidenced by the early translation of the *Zīj al-Sindhind*. The calendars produced in it were transformed to the Arabic lunar calendar (*ʿala Sinnī al-ʿArab*), which would not have been possible without significant knowledge of mathematics. Saliba, *Al-Fikr al-ʿIlmī*, 81, 83.

37. Gutas, *Greek Thought, Arabic Culture*, 11–16, 116. This, according to Gutas, "would explain the appearance, almost overnight, it would seem, of numerous experts in the court of the ʿAbbāsids once they made the political decision to focus the efforts of the available scientists and sponsor the translation of written sources" (16).

38. In the field of agronomy, for example, the ninth-century botanical lexicon *Kitāb al-Nabāt* (The Book of Plants) by Abū Ḥanīfa al-Dīnawarī (d. 895) represents the culmination of a tradition in which autonomous botanical writings of a quasi-scientific nature were part of the sciences written

about in the Arabic language. Similar works were composed on animals and other natural phenomena. Another work, *Kitāb al-Anwā'* by Ibn Quṭayba al-Dīnawarī (d. 889), is a lexicon of the knowledge known to the Arabs before Islam about the heavens, stars, planetary risings and settings, stations of the sun and the moon, and the connection of all of these to weather changes during the different cycles of the solar year. On *Kitāb al-Anwā'* see Saliba, *Al-Fikr al-'Ilmī*, 75–76.

39. Al-Khalīl bin Aḥmad al-Farāhīdī, *Kitāb al-'Ayn*, quoted in Rashed, *Tārīkh al-Riyādiyyāt al-'Arabiyya*, 294–98. In Arabic names, *ibn* and *bin* are interchangeable.

40. For example, the imperial ideological context of the early 'Abbāsids, as well as the centralizing tendencies under the 'Abbāsid caliph Al-Ma'mūn, both described by Gutas (*Greek Thought, Arabic Culture*, 28–104), were abandoned by the time of the caliph Al-Mutawakkil, and the Mu'tazilī theology adopted by Al-Ma'mūn receded in the interest of traditionalism; still, translations and scientific activities continued unabated under Al-Mutawakkil, who was the patron of the most famous of the translators, Ḥunayn bin Isḥāq. See Saliba, *Al-Fikr al-'Ilmī*, 26. In any case, the practical needs of the eighth and early ninth centuries were different from those of later centuries.

41. See, for example, Edward Grant, *The Foundations of Modern Science in the Middle Ages: Their Religious, Institutional and Intellectual Contexts* (Cambridge: Cambridge University Press, 1997), 185; and Toby E. Huff, *The Rise of Early Modern Science: Islam, China and the West* (New York: Cambridge University Press, 1993). For a critique of Huff see George Saliba, review of Toby E. Huff's *The Rise of Early Modern Science: Islam, China and the West* (rights reserved), *BRIIFS* 1, no. 2 (1999). Also see the response by Huff, "The Rise of Early Modern Science: A Reply to George Saliba," *BRIIFS* 4, no. 2 (2002); and the subsequent response by George Saliba, "Flying Goats and Other Obsessions: A Response to Toby Huff's Reply," *BRIIFS* 4, no. 2 (2002). Another common assertion is the one noted earlier regarding the conflict between science and religion. I shall return to this topic in the third chapter.

42. "Neither the *madrasa* nor its cognate institutions harbored any but the religious sciences and their ancillary subjects. If such was the case, how is one to explain the flourishing of the philosophical and natural sciences?" George Makdisi, *The Rise of Colleges: Institutions of Learning in Islam and the West* (Edinburgh: Edinburgh University Press, 1981), 77.

43. "The introduction of Greek works into Islam had a profound influence on the development of Islamic thought and education. Islam . . . had to

face the problem of how to assimilate the 'pagan' knowledge of the Greeks to a conception of the world that included God as its creator. The development of Islamic thought that attempted to bring a solution to this problem took place both within and without institutionalized learning. The solution, such as it was, came as a result of the interplay between the traditionalist forces represented by the *madrasa* and cognate institutions, and rationalist forces represented by *dār al-'ilm* and its cognates. By the time the traditionalist institutions had won the battle against those of rationalism and absorbed them, they had also absorbed a great amount of what they had originally opposed." Makdisi, *Rise of Colleges*, 77–78.

44. Makdisi, *Rise of Colleges*, 77–78.

45. For critiques of Makdisi see Jonathan Berkey, *The Transmission of Knowledge in Medieval Cairo: A Social History of Islamic Education* (Princeton, NJ: Princeton University Press, 1992), 7, 14–16, 22, 43–44; Michael Chamberlain, *Knowledge and Social Practice in Medieval Damascus, 1190–1350* (Cambridge: Cambridge University Press, 1994); Daphna Ephrat, *A Learned Society in a Period of Transition: The Sunni 'Ulama' of Eleventh-Century Baghdad* (Albany: State University of New York Press, 2000). Devin Stewart provides a compelling response to these critiques, specifically in connection to the question of formally issuing licenses (*ijāzas*); he demonstrates the "regular granting of the licenses to issue legal opinions and/or teach law in Mamlūk territory throughout the thirteenth and fourteenth centuries." Devin Stewart, "The Doctorate of Islamic Law in Mamlūk Egypt and Syria," in Joseph Lowry, Devin Stewart, and Shawkat Toorawa, eds., *Law and Education in Medieval Islam: Studies in Memory of Professor George Makdisi* (Cambridge, UK: Gibb Memorial Trust, 2004), 45–90. However, Stewart does not address the issue of education in the rational sciences.

46. Sonja Brentjes, "On the Location of the Ancient or 'Rational' Sciences in Muslim Educational Landscapes (AH 500–1100)," *Bulletin of the Royal Institute of Inter-Faith Studies* 4 (2002): 47–71.

47. Numerous medieval Muslim scholars specialized in both the rational and religious sciences, and although many historians have noted the multiple specializations by individual scholars, there is very little detailed study of the actual interaction of religious and scientific knowledge in the scholars' works. Two recent exceptions are worth mentioning. The first is the work of Robert Morrison, who analyzes the complex relationship between the religious and the scientific scholarship of Niẓām al-Dīn al-Nīshābūrī (d. c. 1330), focusing on his Qur'ānic exegesis and his astronomy. See Robert Morrison, "The Por-

trayal of Nature in a Medieval Qur'ān Commentary," *Studia Islamica* (2002): 1–23; see also Morrison's forthcoming intellectual biography of Al-Nīshābūrī. The other exception is the excellent recent dissertation by Nahyan Fancy, "Pulmonary Transit and Bodily Resurrection," in which he examines the relationship between the religious, philosophical, and scientific thought of Ibn al-Nafīs, a specialist in ḥadīth and medicine who wrote on philosophy. Fancy brilliantly explains the discovery of the pulmonary transit of the blood in this context.

48. In the twelfth to fourteenth centuries many scholars combined their interests in ḥadīth and medicine—Ibn al-Nafīs, for one (see previous note). Another example is the Egyptian scholar 'Izz al-Dīn Ibn Jamā'a (d. AH 819), who had profound knowledge of medicine and surgery and composed works on ḥadīth and *kalām* (speculative theology). Ibn Jamā'a is described as the "Shaykh of Egypt in the rational sciences" and as the "miracle of miracles in the knowledge of the literary sciences, the rational sciences, and the two *aṣl*s [i.e., the principles of religion and the principles of jurisprudence]." The combination of the two aṣls and logic was also very common, as with the famous Sāhfi'ī scholar Taqī al-Dīn al-Subkī (d. AH 786), the Ash'arī theologian Shams al-Dīn al-Iṣfahānī (d. AH 688), and the Qur'ān commentator Al-Bayḍāwī (d. AH 691). Brentjes, "On the Location of the Sciences," 54–55. Ibn al-Bayṭār (d. 1248), one of the leading physicians and botanists of his time, was also a leading jurist who collaborated in his medical research with a circle of Syrian and Egyptian Ḥanbalī scholars. In Cairo this circle included Abū al-Faraj 'Abd al-Laṭīf ibn 'Abd al-Mun'im al-Ḥarrānī (13th c.), the leading Ḥanbalī jurist of the time. For these and other examples of men who combined religious scholarship and scholarship in the rational sciences, especially the mathematical ones, see 'Abd al-Qādir Muḥammad al-Nu'aymī al-Dimashqī, *Al-Dāris fī Tarīkh al-Madāris* (Beirut: Dār al-Kutub al-'Ilmiyya, 1990), 1:58, 66, 67, 172, 186, 193, 320; 2:28.

49. One scholar who examined both is Al-Nīshābūrī, mentioned above.

50. Ṣadr al-Sharī'a al-Bukhārī (d. 1347), for example, wrote on such traditional Islamic disciplines as Arabic grammar, rhetoric, theology, legal stipulations and contracts, and Ḥanafī law and was one of the leading religious authorities of the Ḥanafī school. In addition to these works on traditional Islamic disciplines, Ṣadr al-Sharī'a wrote a three-volume encyclopedic survey, *Ta'dīl al-'Ulūm* (The Adjustment of the Sciences), in the style of commentaries. In it Ṣadr al-Sharī'a presents his assessment of logic, theology, and astronomy and

then attempts to contribute to each field. In the book on astronomy, Ṣadr al-Sharīʿa continues an ongoing tradition of reforming Ptolemaic astronomy and comments in particular on the celebrated *Al-Tadhkira* of Naṣīr al-Dīn al-Ṭūsī (d. 1274) and on *Al-Tuḥfa al-Shāhiyya* by Quṭb al-Dīn al-Shīrāzī (d. 1311) and provides alternative solutions for some of the problems that these two scholars did not tackle. On Ṣadr al-Sharīʿa see Ahmad Dallal, *An Islamic Response to Greek Astronomy: The Astronomical Work of Ṣadr al-Sharīʿa al-Bukhārī* (Leiden: Brill, 1995). On the work of Al-Ṭūsī see *Naṣīr al-Dīn al-Ṭūsī's Memoir on Astronomy: Al-Tadhkira fī ʿIlm al-Hayʾa*, ed. and trans. Jamil Ragep (New York: Springer-Verlag, 1993). On the work of Al-Shīrāzī see Kennedy, *Studies in the Islamic Exact Sciences*, 84–97.

51. For example, ʿAbd al-Laṭīf al-Baghdādī (d. 1231) says that he used to teach religious disciplines in the Azhar mosque in Cairo from morning till about four in the afternoon; then students could come to study medicine and other subjects with him until the end of the day, at which time he would go back to the mosque to teach some more. Quoted in Ibn Abī Uṣaybiʿa, *ʿUyūn al-Anbāʾ fī Tabaqāt al-Aṭṭibāʾ* (Beirut: Maktabat al-Ḥayāt, n.d.), 689.

52. In Arabic the main words used to classify the sciences are *taṣnīf* (categorization), *marātib* (levels), *tartīb* (ordering), and *taqṣīm al-ʿulūm* (division of sciences). The presence of the rational sciences in various classifications reflects not just a social but also an epistemological sanction of these sciences.

53. Some scholars have suggested that this alliance "circumscribed the directions into which these sciences led to and cut off routes that they might have followed under other configurations." Brentjes, "On the Location of the Sciences," 53; see also the arguments by A. I. Sabra. However, as the qibla example at the beginning of this chapter illustrates, these sciences were integrated into the religious curriculum but retained their cognitive autonomy. There is compelling evidence that this integration or partnership enhanced the standing of science and boosted independent scientific activity rather than inhibited it. With or without this partnership, most scientific research was pursued in areas that had nothing to do with the practical needs of religion.

54. A typical description is "He was the leading scholar of his time in the rational sciences." Among the more famous scholars of the rational sciences mentioned in the *ṭabaqāt* literature and biographical dictionaries are Abū Naṣr al-Farābī (d. 950), Ibn Sīnā (d. 1037), ʿAbd al-Laṭīf al-Baghdādī (d. 1231), Naṣīr al-Dīn al-Ṭūsī (d. 1274), Quṭb al-Dīn al-Shīrāzī (d. 1311), Saʿd al-Dīn al-Taftazānī (d. 1390), ʿAḍuḍ al-Dīn al-Ījī (d. after 1355), Al-Sharīf al-Jurjānī

(d. 1413), Fakhr al-Dīn al-Rāzī (d. 1209/1210). The last is described by Ibn Qāḍī Shuhba as the chief imam dealing with the rational sciences and one of the imams dealing with the sciences of *sharī'a* (Islamic law). See Brentjes, "On the Location of the Sciences," 57, 70, and passim.

55. On the function of biographical dictionaries see Wadad al-Qadi, "Biographical Dictionaries," in Gerhard Endress, ed., *Organizing Knowledge: Encyclopaedic Activities in the Pre-Eighteenth Century Islamic World* (Leiden: Brill, 2006), 23–75.

56. Ibn Abī Uṣaybi'a, *'Uyūn al-Anbā fī Tabaqāt al-Aṭṭibā'*.

57. On hospitals see 'Īsā, *Tārīkh al-Bīmāristānāt fī al-Islām*. On endowed medical schools see Al-Nu'aymī, *Al-Dāris fī Tārīkh al-Madāris*, 2:100 ff.; and Ibn Abī Uṣaybi'a, *'Uyūn al-Anbā fī Tabaqāt al-Aṭṭibā'*, 728 ff.

58. For example, 'ilm al-mīqāt (timekeeping) was taught at the Sultan Ḥasan madrasa, established in the second half of the fourteenth century, as outlined in the foundational deed of the school, although the majority of its students studied Islamic law. See Brentjes, "On the Location of the Sciences," 60, following Berkey, *Transmission of Knowledge in Medieval Cairo*, 69. On the teaching of inheritance algebra and other mathematical sciences in religious schools see Al-Nu'aymī, *Al-Dāris fī Tārīkh al-Madāris*, 1:56–58, 66.

59. Brentjes notes, on the authority of the historian Al-Sakhāwī, that Shihāb al-Dīn ibn al-Majdī (14th c.) was the timekeeper (*muwaqqit*) of Al-Azhar mosque and head of the teachers at the Madrasa al-Jānibākiyya al-Dawādāriyya, presumably a school devoted to the religious sciences. Al-Majdī wrote three mathematical treatises and twenty-one astronomical treatises and taught arithmetic, algebra, astronomy, timekeeping, astronomical instruments, and philosophy to more than twenty students. Brentjes, "On the Location of the Sciences," 57–58, 60.

60. On observatories and observations see Aydin Sayili, *The Observatory in Islam* (Ankara: Türk Tarih Kurumu, 1960); Muammer Dizer, ed., *International Symposium on the Observatories in Islam, 19–23 September 1977* (Istanbul, 1980). Several studies on observational instruments and the related subject of timekeeping are collected in David King, *Astronomy in the Service of Islam*; King, *Islamic Mathematical Astronomy*; and King, *Islamic Mathematical Instruments*. On early observational activities see Aydin Sayili, "The Introductory Section of Habash's Astronomical Tables Known as the 'Damascene' Zīj," in *Ankara Üniversitesi Dil ve Tarih-Gografya Fakültesi Dergisi* 13, no. 4 (1955). See also Ibn Yūnus, *Kitāb al-Zīj al-Ḥākimī al-Kabīr*, published under the title *Le livre de la Grande Table Hakémite*, ed. and trans. Caussin de Perceval (Paris:

Bibliothèque nationale, 1804). Caussin also reproduces useful quotations from a variety of sources on observational activities in eleventh-century Cairo; especially useful is the detailed account by the historian Al-Maqrīzī on the Afḍal-Baṭā'iḥī observatory. See also Abū al-Rayḥān Muḥammad ibn Aḥmad al-Bīrūnī, *Taḥdīd Nihāyāt al-Amākin li-Tasdīd Masāfāt al-Masākin*, ed. P. G. Bulgakov (Cairo: Revue de l'Institut des Manuscrits Arabes, 1962). For an English translation see Jamil Ali, *The Determination of the Coordinates of Cities: Al-Bīrūnī's Taḥdīd al-Amākin* (Beirut: AUB Press, 1967).

On the Marāghā observatory see Faḍl Allāh Rashīd al-Dīn, *Jāmiʿ al-Tawārīkh*, ed. and trans. E. Quatremère (Paris, 1836); Salāḥ al-Dīn Khalīl Aybak al-Ṣafadī, *Al-Wāfī bil-Wafiyyāt*, ed. H. Ritter (Istanbul: Staatsdruckerise, 1931), vol. 1; and Ibn Shākir al-Kutubī, *Fawt al-Wafiyyāt* (Cairo, AH 1299), vol. 2. For a detailed account of the design and use of various instruments at Marāghā see Mu'ayyad al-Dīn al-ʿUrḍī, *Risāla fī Kayfiyyat al-Arṣād*, published under the title "Al-Urdi'nin 'Risalet-ün Fi Keyfiyet-il-Ersad' Adli Makalesi" (in English and Turkish), ed. and trans. Sevim Tekeli, in *Arastirma* 8 (1970). For an earlier German study and translation of Al-ʿUrdī's work see Hugo J. Seemann, "Die Instrumente der Sternwarte zu Maragha nach den Mitteilungen von al-ʿUrdi," in *Sitzungsberichte der physikalisch-medizinischen Sozietät zu Erlangen* 60 (1928): 15–126. On Marāghā in general see André Godard, *Les monuments de Maragha* (Paris: E. Leroux, 1934).

On the instruments of the Samarqand observatory, and also for additional information on the Marāghā instruments, see E. S. Kennedy, trans., "Al-Kāshī's Treatise on Astronomical Observational Instruments," in Kennedy et al., *Studies in the Islamic Exact Sciences*, 394–404. On the career of the same astronomer, his experiences, and his relationship to his patron and colleagues see Kennedy, trans., "A Letter of Jamshīd al-Kāshī to His Father: Scientific Research and Personalities at a Fifteenth Century Court," also in Kennedy et al., *Studies in the Islamic Exact Sciences*, 722–44.

61. The most important correction introduced at these observatories was to show that the apogee of the solar orb (the point on the solar orbit farthest from Earth) moves with the precession of the fixed stars (the slow shifting of the position of the fixed stars). Another important contribution was the determination of the obliquity of the ecliptic (the angle between the plane of the ecliptic and the plane of Earth's equator), one of the earliest and most frequent observational activities to occupy astronomers since the observations conducted at Shammāsiyya.

62. Some of the measurements taken at the Ma'mūn observatories were independently confirmed by private observations right after the death of Ma'mūn, only to be revised later on by other astronomers. Ma'mūn's astronomers, for example, corrected the ancient value for the precession of the equinoxes (the slow westward shift of the equinoxes along the plane of the ecliptic) from one degree every hundred years to one degree every sixty-six years. This value was confirmed in the ninth century in the private observations of the Banū Mūsā brothers, Ḥabash al-Ḥāsib, Al-Nayrīzī, and Al-Māhānī. In the next century, Banū Amājūr (885–933) and Ibn al-A'lam (c. 975) came up with a value of one degree every seventy years. The subject was still being pursued in the thirteenth century when the Marāghā observatory was established.

63. Toward the end of the tenth century, for example, the astronomer Al-Khujandī designed and built in Rayy, under the patronage of the Buwayhid ruler Fakhr al-Dawla, a huge sextant with a radius of twenty meters. A degree of arc on this sextant measured about thirty-five centimeters, enabling much more accurate solar observations. Another large instrument is described by Ibn Sīnā, and a modified version of this instrument was later used in Marāghā.

64. Carlo Nallino, *'Ilm al-Falak, Tārīkhuhu 'ind al-'Arab fī al-Qurūn al-Wusṭā* (Rome: Libro Moderno, 1911).

65. For example, Otto Neugebauer suggests in a passing remark that a Babylonian tradition might have informed the algebra of Al-Khwārizmī, which clearly did not have Greek antecedents. Otto Neugebauer, *The Exact Sciences in Antiquity* (New York: Dover, 1969).

66. See Rashed, *Tārīkh al-Riyādiyyāt*, 19–33, and passim; and Rashed, *Development of Arabic Mathematics.*

67. After Al-Karajī, the central efforts in algebraic research focused on the arithmetization of algebra, a genre of research that was new in both its contents and its organization. For six centuries, and into the seventeenth century, most of the important mathematicians continued to build on and develop the work of Al-Karajī. The work of Al-Samaw'al al-Maghribī (12th c.) is of particular importance. In his *Al-Bāhir fī al-Jabr*, Al-Samaw'al defines algebraic power, studies arithmetical operations on polynomials, and examines the multiplication, division, addition, subtraction, and extraction of roots for irrational quantities. Rashed, *Tārīkh al-Riyādiyyāt*, 33–101.

68. On some of the earliest astronomical works in Arabic see David Pingree, "The Fragments of the Works of Ya'qub b. Tariq," *Journal of Near Eastern Studies* 27 (1968): 97–125; and Pingree, "The Fragments of the Works of al-

Fazari," *Journal of Near Eastern Studies* 29 (1970): 103–23. See also Pingree, "The Greek Influence on Early Islamic Mathematical Astronomy," *Journal of the American Oriental Society* 93 (1973): 32–43.

69. See, for example, Ullmann, *Islamic Medicine;* Michael Dols, introduction to 'Ali Ibn Riḍwān, *Medieval Islamic Medicine;* Conrad, "Arab-Islamic Medical Tradition"; and Pormann and Savage-Smith, *Medieval Islamic Medicine.* See also this useful collection of essays: Meyerhof, *Studies in Medieval Arabic Medicine.*

70. See Ptolemy, *Ptolemy's Almagest,* trans. G. J. Toomer (New York: Springer-Verlag, 1984). The best study of the *Almagest* is Olaf Pedersen, *A Survey of the Almagest* (Odense: Odense University Press, 1974). The first chapter, "The Almagest through the Ages," includes a section on the *Almagest* among the Arabs. See, in particular, pp. 11–25.

71. For a brief overview of early developments in the field of astronomy see Régis Morelon, "General Survey of Arabic Astronomy," in Rashed with Morelon, *Encyclopedia of the History of Arabic Science,* 1:1–19.

72. For editions, translations, and analysis of Thābit's extant astronomical works see Régis Morelon, *Thābit ibn Qurra: Oeuvres d'astronomie,* Collection Science et Philosophie Arabes: Textes et études (Paris: Sociéte d'édition "Les Belles Lettres," 1987). Also see Otto Neugebauer, "Thābit ben Qurra," "On the Solar Year," and "On the Motion of the Eighth Sphere," *Proceedings of the American Philosophical Society* 106 (1962): 264–99; George Saliba, "Early Arabic Critique of Ptolemaic Cosmology: A Ninth-Century Text on the Motion of the Celestial Spheres," *Journal for the History of Astronomy* 25 (1994): 115–41.

73. Abū al-Rayḥān Muḥammad ibn Aḥmad al-Bīrūnī, *Al-Qānūn al-Mas'ūdī,* 3 vols. (Haidarabad: Dā'irat al-Ma'ārif, 1954). Al-Bīrūnī was born in Khwārizm and died in Ghazna. His native tongue was Persian, but he composed the vast majority of his works in Arabic. He also knew Sanskrit and, as a result, achieved full command of Indian astronomy, in addition to the well-established Greek and Arabic traditions.

74. For general surveys of the history of Arabic medicine see, for example, Conrad, "Arab-Islamic Medical Tradition"; and Pormann and Savage-Smith, *Medieval Islamic Medicine.*

75. For Arabic optics see Ibn al-Haytham, *Optics of Ibn al-Haytham;* and Roshdi Rashed, *Geometry and Dioptrics in Classical Islam* (London: Al-Furqan Islamic Heritage Foundation, 2005).

76. See Rashed, *Development of Arabic Mathematics.*

77. Already Abū Naṣr al-Fārābī (d. 950) in his *Iḥṣā' al-'Ulūm* (Classification of the Sciences), ed. Uthmān Amīn (Cairo: Maktabat al-Anglo al-Maṣriyya, 1968), 88–89, includes the name of a new science, *'ilm al-athqāl* (the science of weights), and divides it into a practical branch and a theoretical one. See, for example, the several excellent studies by Mohammed Abattouy, including "The Arabic Science of Weights: A Report on an Ongoing Research Project," *BRIIFS* 1, no. 4 (2000): 109–30; and Abattouy, "*'Ulūm al-Mikānīkā fī al-Gharb al-Islāmī al-Wasīṭ, Dirāsa Awwaliyya,*" in Bennacer El Bouazzati, ed., *La pensée scientifique au Maghreb: Le haut Moyen Age* (Rabat: Mohammed V University Press, 2003); see also the bibliographies of the essays for additional articles by Abattouy. Compare Elaheh Kheirandish, "Organizing Scientific Knowledge: The 'Mixed' Sciences in Early Classifications," in Endress, *Organizing Knowledge,* 135–54. Kheirandish pays special attention to optics and mechanics but does not see much conceptual innovation in Al-Khāzinī's science of weights.

78. Parallel trends in astronomy will be addressed in the following chapters.

79. On Arabic pharmacology see the works by Ibrahim Ibn Murad, especially *Buḥūth fī Tārīkh al-Ṭibb wal-Ṣaydala 'ind al-'Arab* (Beirut: Dār al-Gharb al-Islāmī, 1991).

80. Plutarch, *Plutarch's Lives,* trans. John Dryden; ed. Arthur Hugh Clough (New York: Random House, 2001), 480 ff. (entry on Marcellus).

81. On technology see Ahmad Y. al-Hasan and Donald Hill, *Islamic Technology: An Illustrated History* (Cambridge: Cambridge University Press, 1986); and Banū (Sons of) Mūsā bin Shākir, *The Book of Ingenious Devices (Kitāb al-Ḥiyal),* trans. Donald Hill (Boston: D. Reidel, 1989).

82. Al-Jazarī, *The Book of Knowledge of Indigenous Mechanical Devices,* trans. Donald Hill (New York: Springer-Verlag, 1973).

83. Ahmed Tahiri, *Agricultura y medicina en al-Andalus: Dimension filosofica y tendencia experimental* (Muhammadiyya: Ḥasan II University Press, 1997), 97–101. Tahiri surveys several works on agriculture; the references here are to a manuscript by Muḥammad b. Mālik al-Taghnarī (11th c.), written during what historians often refer to as the Andalusian agricultural revolution. For more on agriculture see Expircion Garcia Sanchez, "Agriculture in Muslim Spain," in Salma Khadra Jayyusi, ed., *The Legacy of Muslim Spain* (Leiden: Brill, 1992), 127–46; Lucie Bolens, *Agronomies Andalous au Moyen Age* (Geneva: Droz, 1981); Lucie Bolens, "La révolution agricole andalouse du XI siècle," *Studia Islamica* 47 (1978): 121–41.

84. A standard work on timekeeping is the famous *Jāmiʿ al-Mabādiʾ wal-Ghayāt fī ʿIlm al-Mīqāt* by Abū ʿAlī al-Marrākushī (Cairo, c. 1280); it includes theoretical treatments of spherical astronomy and sundial theory, discussions of the construction and use of various instruments, and extensive tables. Tables for various locations were often compiled in connection with timekeeping, as were auxiliary trigonometric tables, which were compiled to facilitate the solution of problems of spherical trigonometry. The tables of the fourteenth-century Damascene timekeeper Al-Khalīlī are examples of the finest accomplishment within this tradition. They were the most accurate and exhaustive numerical solutions for all timekeeping problems and for the direction of the qibla. See the studies by David King cited above in notes 23, 24, and 60.

85. Perhaps at the expense of the traditional unity between science and philosophy.

86. See Ahmad Dallal, "Ibn al-Haytham's Universal Solution for Finding the Direction of the *Qibla* by Calculation," *Arabic Sciences and Philosophy* 5 (1995): 145–93.

87. As, for example, in Grant, *Foundations of Modern Science;* and Huff, *Rise of Early Modern Science.* See also David Lindberg, *The Beginnings of Western Science* (Chicago: Chicago University Press, 1992), 161–82.

88. For example, some say that although Thābit Ibn Qurra proved the theorem of Fermat, his proof was forgotten, and therefore it could not have had an influence on Fermat's later proof (17th c.). Rashed shows that there are continuous references to Thābit's proof, mostly in works dedicated to teaching, including works by Al-Qabīsī (10th c.), Al-Karajī (late 10th c.), Al-Baghdādī (d. 1231), Kamāl al-Dīn al-Fārisī (d. 1319), and Al-Tanūkhī (14th c.). Rashed, *Tārīkh al-Riyādiyyāt,* 299 ff., 307–11.

In medicine, much of the earlier research on Ibn al-Nafīs was aimed at characterizing his discovery of the pulmonary transit of the blood as a happy guess that had no influence on the "scientific" discovery of the pulmonary circulation of the blood by Harvey. See, for example, Max Meyerhof, "Ibn al-Nafīs and His Theory of the Lesser Circulation," *Isis* 23 (1935): 100–120. Recent scholarship has shown that Ibn al-Nafīs's discovery had a distinct influence on later European anatomical theories of blood circulation, despite significant differences in the theories. More important, Nahyan Fancy situates Ibn al-Nafīs's scientific theories within the context of his own intellectual milieu, and especially in the context of the physiological and philosophical questions of his time. Fancy, "Pulmonary Transit and Bodily Resurrection."

In optics, to give one more example, it has often been argued that Ibn al-Haytham's (d. 1039) work had no effect on Arabic optical research and was appreciated only in Latin European scholarship. Yet the work of Al-Fārisī, which builds on Ibn al-Haytham's work, and recent evidence for an eleventh-century Andalusian version of Ibn al-Haytham's *Manāẓir* seem to illustrate a continuity in creative optical research not just before and leading to Ibn al-Haytham but also after him.

89. Al-Bīrūnī also collaborated with another astronomer to simultaneously observe the same astronomical phenomenon in two different locations in order to measure the distance between them.

90. On this research tradition see George Saliba, *A History of Arabic Astronomy: Planetary Theories during the Golden Age of Islam* (New York: New York University Press, 1995).

91. See Rashed, *Development of Arabic Mathematics;* Roshdi Rashed, *Sharaf al-Dīn al-Ṭūsī: Oeuvres mathématiques: Algèbre et géométrie au XIIe siècle* (Paris: Sociéte d'édition "Les Belles Lettres," 1986). Rashed traces a series of renewals and new beginnings for the science of algebra, each building on earlier traditions but developing the field in a new direction. The first beginning, as it were, was the establishment of the science of algebra (not just algebraic operations) by Al-Khwārizmī. A second beginning is attributed to Al-Karajī, who formulated a deliberate project to arithmetize algebra and reshaped algebra as the science of computing unknowns, expanding in the process the concept of numbers (previously, algebra dealt with the computation of specific entities). A third beginning, undertaken by Al-Khayyām, built on a possibility suggested in the research of Al-Karajī and his school: applying algebra fruitfully to other fields. Mathematicians before Al-Karajī extracted square and cubic roots but lacked an abstract algebraic arithmetic, so they could not generalize their algorithms. Because of the difficulty of providing straightforward algebraic solutions for third degree equations, Al-Khayyām used geometry to solve them. In this phase, therefore, algebra was advanced by means of geometry; at a later stage, curvatures were studied by means of their equations, an approach that laid the foundations of algebraic geometry. See Rashed, *Tārīkh al-Riyādiyyāt,* 70, 161–63, 183–84, 364.

Chapter 2. Science and Philosophy

1. See, for example, R. Arnaldez, "Comment c'est ankylosée la pensée philosophique dans l'Islam?" in R. Brunschvig and G. E. von Grunebaum,

eds., *Classicisme et déclin culturel dans l'histoire de l'Islam* (Paris: G. P. Maisonneuve et Larose, 1977), 247–59. The formulation of this view is traceable to the works of the French Orientalist Ernest Renan, who subscribed to the positivist Comtean view that real science explores the secret causes behind the natural phenomena. Interestingly, like many advocates of this positivist view of science, Renan was a philosopher, not scientist, and he was wedded to theories about science that were outdated even by the standards of his own times.

2. For an example of the interest in Islamic science merely on account of its preservation of the Greek scientific legacy see David Lindberg, *The Beginnings of Western Science* (Chicago: University of Chicago Press, 1992). Lindberg's chapter on "Science in Islam" (161–82) is divided into subsections with the following titles: "Learning and Science in Byzantium," "The Eastward Diffusion of Greek Science," "The Birth, Expansion, and Hellenization of Islam," "Translation of Greek Science into Arabic," "The Islamic Response to Greek Science," "The Islamic Scientific Achievement," and "The Decline of Islamic Science."

3. See, for example, C. H. Becker, "*Turāth al-Awā'il fī al-Sharq wal-Gharb*," in 'Abd al-Raḥmān Badawī, ed., *Al-Turāth al-Yūnānī fī al-Ḥadāra al-Islāmiyya* (Beirut: Dār al-Qalam, 1980), 3–33. Becker's essay is a translation of his 1931 "Das Erbeder Antike in Orient und Okzident." For later formulations of similar views, and the argument that Islamic cultural imperatives militated against the development of science and eventually led to its decline, see G. E. von Grunebaum, *Islam: Essays in the Nature and Growth of a Cultural Tradition* (London: Routledge and Kegan Paul, 1961), especially chap. 6; and F. E. Peters, *Aristotle and the Arabs: The Aristotelian Tradition in Islam* (New York: New York University Press, 1968), chap. 4.

4. For example, Edward Kennedy maintains that "it is well to stress at the outset that the impulse behind the activity we describe was theoretical, and in some sense philosophical, rather than an attempt to improve the bases of practical astronomy." E. S. Kennedy, "Late Medieval Planetary Theory," in E. S. Kennedy, colleagues, and former students, *Studies in the Islamic Exact Sciences*, ed. David King and Mary Hellen Kennedy (Beirut: American University of Beirut, 1983), 85. Although Kennedy does not mean here to question the scientific value of the Islamic contributions in planetary theory, other historians of science derive such conclusions from his observation.

5. See, for example, Pierre Duhem (d. 1916), *To Save the Phenomena: An Essay on the Idea of Physical Theory from Plato to Galileo*, trans. Edmund

Doland and Chaninah Maschler (Chicago: University of Chicago Press, 1969).

6. That the new sciences emerged because of the mathematization of nature is one of many theories explaining modern science and the scientific revolution.

7. I shall give corroborating examples from other disciplines before venturing some tentative generalizations.

8. Otto Neugebauer, *The Exact Sciences in Antiquity* (New York: Dover, 1969), 2.

9. For a brief overview of early developments in astronomy see A. Aaboe, "Scientific Astronomy in Antiquity," *Philosophical Transactions of the Royal Society of London*, series A, 276 (1974): 21–42.

10. For a standard English version see Ptolemy, *Ptolemy's Almagest*, trans. G. J. Toomer (New York: Springer-Verlag, 1984).

11. Régis Morelon, "La version arabe du livre des hypothèses de Ptolémée," *MIDEO* (Periodical of the Dominican Institute for Oriental Studies) 21 (1993): 7–85. *The Planetary Hypothesis* of Ptolemy is extant only in its Arabic translation. Since it had no Latin translation in the Middle Ages, the only possible source for medieval discussions of it in Europe must have been based directly on Arabic sources.

12. For an excellent concise exposition of Aristotelian celestial physics see S. Sambursky, *The Physical World of Late Antiquity* (Princeton, NJ: Princeton University Press, 1987), 122–53, especially 133–45, on Ptolemy.

13. Aristotle's main treatment of the heavenly region is in *De Caelo* (On the Heavens); he also addressed it to some extent in his *Physics* and his *Metaphysics*. He discusses the sublunar region in *On Generation and Corruption*, in books 3 and 4 of *De Caelo*, and in *Meteorology*.

14. Aristotle, *Physics*, in *The Complete Works of Aristotle: The Revised Oxford Translation*, ed. Jonathan Barnes (Princeton, NJ: Princeton University Press, 1984), 3341–42.

15. Were vacuum to be allowed, a compulsory motion in vacuum would not have to be a motion between two "natural" places. A rock thrown upward, away from the center of the Earth, could potentially have an indefinite motion and would not stop unless an outside force acted on it. But indefinite motion in the sublunar region is an impossibility in Aristotelian natural philosophy; it follows that vacuum is an impossibility in his cosmology.

16. The best concise account of the post-eleventh-century develop-

ments in planetary theory is given in George Saliba, "Arabic Planetary Theories after the Eleventh Century AD," in Roshdi Rashed with Régis Morelon, eds., *Encyclopedia of the History of Arabic Science*, 3 vols. (London: Routledge, 1996), 1:58–127. For more detailed accounts see George Saliba, *A History of Arabic Astronomy: Planetary Theories during the Golden Age of Islam* (New York: New York University Press, 1944); and George Saliba, *Al-Fikr al-'Ilmī al-'Arabī; Nash'atuhu wa Taṭawwuruhu* (Balamand, Lebanon: Balamand University, 1998).

17. The essays by Saliba cited in the previous note contain detailed descriptions of the Ptolemaic models.

18. The name Marāghā school is often given to the eastern reformers in recognition of the achievements of a number of astronomers working in an observatory established at Marāghā. Although the contributions of these astronomers are no doubt monumental, the reform of Ptolemaic astronomy started before the establishment of the Marāghā observatory in the thirteenth century and reached its highest point in the fourteenth. Some of the astronomers of the Marāghā group started their reform projects before they joined this observatory; they were perhaps invited to join the observatory team because they were already engaged in pertinent research. This may be true of Al-Ṭūsī and Al-'Urḍī, both of whom commenced their reformative work before the establishment of the observatory. The eastern reform tradition, then, was too diffuse to be associated with any one geographical area or period; rather, it describes several centuries of Arabic astronomical research throughout the eastern domains of the Muslim world.

On the astronomical work of Al-'Urḍī see Mu'ayyad al-Dīn al-'Urḍī, *Kitāb al-Hay'ah: The Astronomical Work of Mu'ayyad al-Dīn al-'Urdī, a Thirteenth Century Reform of Ptolemaic Astronomy*, ed. George Saliba (Beirut, 1990), 31 f. On the work of Al-Ṭūsī see *Naṣīr al-Dīn al-Ṭūsī's Memoir on Astronomy: Al-Tadhkira fī 'Ilm al-Hay'a*, ed. and trans. Jamil Ragep (New York: Springer-Verlag, 1993), 65 f.

19. For a detailed account of the various models proposed by the astronomers of the eastern parts of the Muslim world to solve the problems of Ptolemaic astronomy see Saliba, "Arabic Planetary Theories," especially 86 ff.

20. On the Ṭūsī couple see Saliba, "Arabic Planetary Theories," 94–95.

21. On the 'Urḍī lemma see Saliba, "Arabic Planetary Theories," 106.

22. Significant scientific activity started in Al-Andalus in the ninth cen-

tury; initially, this activity was almost completely dependent upon and lagging behind the sciences of the eastern part of the Muslim world. Between the ninth and eleventh centuries, however, a full-fledged scientific tradition emerged. Many scientists traveled east to study science; scientific books were systematically acquired, and large private and public libraries were established. A solid familiarity with the eastern astronomical tradition led, in the eleventh century, to intensive and at times original astronomical activity there. The main astronomers of this period were Maslama al-Majrīṭī of Córdoba and his student Ibn al-Ṣaffār, as well as Al-Zarqīyāl (Zarqallu). Al-Zarqīyāl was one of the main contributors to the compilation of the celebrated *Toledan Tables*, which exerted great influence on the development of astronomy in Latin Europe. These and other astronomers focused on the compilation of tables and on spherical astronomy. They made some new observations, but their primary original contributions mostly had to do with the mathematics of the trepidation movement of the stars and the invention of highly sophisticated astronomical instruments. During this entire period, little work of significance was devoted to planetary theory. For a collection of essays on astronomy in Islamic Spain and North Africa see Julio Samso, *Islamic Astronomy and Medieval Spain* (Aldershot, UK: Variorum, 1994), especially chaps. 1, 8, 9, and 19.

23. For a brief outline of the proposals of the astronomer-philosophers of the western parts of the Muslim world see Saliba, "Arabic Planetary Theories," 84–86. For a fuller account of the philosophical rationale behind the proposed models see A. I. Sabra, "The Andalusian Revolt against Ptolemaic Astronomy: Averroes and al-Bitrūjī," in E. Mendelsohn, ed., *Transformation and Tradition in the Sciences* (Cambridge: Cambridge University Press, 1984), 133–53.

24. A common, perhaps prevalent view in contemporary scholarship attributes the steady decline of the intellectual sciences in Al-Andalus and North Africa to the rise of the so-called fundamentalist states of the Almoravids (Al-Murābiṭūn, 1091–1144) and Almohads (Al-Muwaḥḥidūn, 1147–1232). It was precisely while these dynasties were in power, however, that the greatest Andalusian philosophers worked under the rulers' patronage. What we have, therefore, is not a steady decline of the intellectual disciplines but the rise of some at the expense of others. The decline of mathematical astronomy has nothing to do with the Almoravids or the Almohads, nor with an alleged theological counterrevolution. Rather, the decline was a result of the adoption of a specific research program of astronomical research, a program driven by

the untenable, and by then outdated, Aristotelian philosophical concerns that proved incompatible with the advanced mathematical and scientific aspects of astronomy.

25. See Saliba, *Al-Fikr al-ʿIlmī,* 92–93, as well as the other books and essays by Saliba, cited above.

26. Ibn al-Haytham, *Al-Shukūk ʿalā Baṭlamyūs,* ed. A. Sabra and Nabil Shehaby (Cairo: Dār al-Kutub Press, 1971); hereafter cited in the text.

27. See George Saliba, *A History of Arabic Astronomy;* Saliba, "Al-Qushjī's Reform of the Ptolemaic Model for Mercury," *Arabic Science and Philosophy* 3 (1993): 161–203; Saliba, "A Sixteenth-Century Arabic Critique of Ptolemaic Astronomy: The Work of Shams al-Dīn al-Khafrī," *Journal for the History of Astronomy* 25 (1994): 15–38; Saliba, "A Redeployment of Mathematics in a Sixteenth-Century Arabic Critique of Ptolemaic Astronomy," in A. Hasnawi, A. Elamrani-Jamal, and M. Aouad, eds., *Perspectives arabes et médiévales sur la tradition scientifique et philosophique grecque. Actes du Colloque de la SIHSPAI: Paris 29 mars–3 avril 1993* (Paris: Peeters, 1997), 105–22; *Naṣīr al-Dīn al-Ṭūsī's Memoir on Astronomy;* and Al-ʿUrḍī, *Kitāb al-Hay'ah.*

28. Ibn al-Haytham's work is a critique not just of the *Almagest* but also of Ptolemy's *Planetary Hypothesis* and *Optics.*

29. Ibn al-Haytham's short critique of Ptolemy's *Optics* is very different from his critique of astronomy; the former is based on Ibn al-Haytham's own research in the field of optics and incorporates empirical evidence and mathematical objections, as well as new conceptions of visual perception. In other words, Ibn al-Haytham's critique of *Optics* is not based simply on philosophical grounds.

30. For example, Ibn al-Haytham, *Al-Shukūk*, 33, 38, and passim.

31. This crisis was perceived in many different ways, but even if the various criticisms did not always produce unified research projects or provide clearly articulated alternative paradigms for the practice of astronomy, they still were expressions of a perceived crisis in the old paradigm. As such, the "doubts" genre triggered a program of astronomical research by identifying the errors in the field that needed correction. In a sense, the history of all science is a history of discovering errors in it, not the history of discovering scientific truths but the history of identifying what is not science, the false conceptions that science needs to overcome. The identification of gaps and errors propels further scientific research.

32. Ibn Sīnā, *Al-Shifā',* vol. 4: *ʿIlm al-Hay'a,* ed. Muḥammad Riḍā Mu-

dawwar and Imām Ibrāhīm Aḥmad (Cairo: Wizarat al-Ma'ārif, 1980), 651. Ibn Sīnā repeats the importance of taking into consideration all the refinements introduced in Arabic astronomy and promises to do further work along those lines (659). It should be noted that Ibn Sīnā contributed in his own way to the doubts tradition; in his *Al-Ḥikma al-Mashriqiyya* (Eastern Philosophy) he sums up the areas of disagreement with Aristotle. The best study of Ibn Sīnā remains Dimitri Gutas, *Avicenna and the Aristotelian Tradition* (Leiden: Brill, 1988).

33. Al-Bīrūnī and Ibn Sīnā, *Al-As'ilah wa'l-Ajwibah (Questions and Answers), Including the Further Answers of al-Bīrūnī and al-Ma'ṣūmī's Defense of Ibn Sīnā*, ed. and with English and Persian introductions by Seyyed Hossein Nasr and Mahdi Mohaghegh (Tehran: High Council of Culture and Art, Centre of Research and Cultural Coordination, AH 1352). This book is henceforth cited in the text. Al-Bīrūnī's *Al-As'ila* anticipated the medieval genre of "catalogues of questions on medieval cosmology," which included commentaries and questions on Aristotle's *De Caelo* and was extremely important in the development of medieval science in Europe. Edward Grant, *Planets, Stars, and Orbs: The Medieval Cosmos, 1200–1687* (Cambridge: Cambridge University Press, 1996), appendixes I and II. See also Jean Buridan's (d. 1358) questions on Aristotle's *De Caelo*, "Quaestiones," in Edward Grant, *A Source Book in Medieval Science* (Cambridge: Harvard University Press, 1974).

34. Abū [al-]Rayḥān [Muḥammad ibn Aḥmad] al-Bīrūnī, *Istī'āb al-Wujūh al-Mumkina fī Ṣan'at al-Asṭurlāb*, Bodleian MS Marsh 701, 267r, Bodleian Library, Oxford University.

35. I do not mean to suggest that philosophy, as it was understood at the time, had no impact on astronomy; it surely did. Rather, I wish to account for Al-Bīrūnī's understanding of his own professional practice and to take this understanding seriously. A similar logic applies to the epistemological distinction that Al-Bīrūnī makes between mathematical astronomy and astrology. It is not the backward projection of distinctions that we make today but an attempt to take seriously scientists' assertions about their differences. Al-Bīrūnī's awareness of a distinction between the science of astronomy and philosophy must have sprung from his own practice of mathematical astronomy, but his awareness of a distinction must, in turn, have informed and guided his practice and possibly the practice of other astronomers.

36. See, for example, questions 11–12, where, in an answer to Al-Bīrūnī's critique of the Aristotelian denial of either the levity or the heaviness of the

heavenly orbs and the assertion that the natural motion of the orbs is circular, Ibn Sīnā provides the proof for these arguments from Aristotle's *Metaphysics*, *Physics*, and *Generation and Corruption*.

37. "Why did Aristotle posit that the orb [heavenly sphere] is neither heavy nor light on account of the lack of movement [of the orb] from or toward the center? We can imagine, though not necessarily posit, that it can be one of the heaviest solids . . . or that it is one of the lightest . . . or the impossibility of void outside it." Al-Bīrūnī and Ibn Sīnā, *Al-As'ilah*, 2.

38. "As for the circular motion [of the orb], it does not have to be natural to it, as are the assumed natural motions of the planets to the east along with their compulsory motions to the west. If someone says that these latter motions are not accidental, since there can be no opposition in circular motions and no difference in their directions, then this is nothing more than deception and sophistry, since we cannot posit two natural motions to the same body." Al-Bīrūnī and Ibn Sīnā, *Al-As'ilah*, 2–3. In the response Ibn Sīnā insists that the circularity of the motion of the heavenly bodies is natural, and provides the proof for it from the *Physics;* he adds that the "motions of orbs are not contradictory but only different" (11–12).

39. "Why was Aristotle repulsed by the view that there could be another world outside the one in which we are, one which has a different nature, because we did not come to know the four temperaments and elements after we encountered them, just as a blind person would not be able to imagine on his own what seeing is without hearing about it from people. . . . Alternatively [the other world] could have the same nature [as our world], but its motions could be compounded differently, and the two worlds could be separated from each other by a partition." Al-Bīrūnī and Ibn Sīnā, *Al-As'ilah*, 19. Ibn Sīnā argues that it is logically impossible to have multiple worlds of the same nature, because their shared nature would unite them; he also argues that it is not possible for another world to exist that not only has a different nature from our world's but also has bodily qualities.

40. See, for example, Al-Bīrūnī and Ibn Sīnā, *Al-As'ilah*, 47, where Al-Bīrūnī uses the example of a bottle from which air has been sucked, which would, if placed over water, suck water in to fill the vacuum created within it. Also see p. 58. In *Al-Kashf 'an Manāhij al-Adilla*, the philosopher Ibn Rushd (d. 1198) rejects the Mu'tazilī view that the nonexistent is a being, because this leads to acceptance of the idea of the void/vacuum. This argument suggests a possible theological background to Al-Bīrūnī's suggestion that vacuum might exist. Abū al-Walīd Muḥammad Ibn Rushd, *Faṣl al-Maqāl wa Taqrīr mā bayna*

al-Sharī'a wal-Ḥikma min Ittiṣāl and *Al-Kashf 'an Manāhij al-Adilla fī 'Aqā'id al-Milla* (Beirut: Dār al-Āfāq al-Jadīda, 1978), 139.

41. In *Al-Qānūn al-Mas'ūdī,* Al-Bīrūnī criticizes Ptolemy for using arguments from natural philosophy to prove the sphericity of the heavens; each discipline, he says, should derive its own principles and not be at the mercy of an outside discipline. Al-Bīrūnī, *Al-Qānūn al-Mas'ūdī,* 3 vols. (Haidarabad: Dā'irat al-Ma'ārif al-'Uthmāniyya, 1954), 1:27. Perhaps an equally important aspect of this criticism is that Al-Bīrūnī's deployment of mathematics in connection with a principle of natural philosophy is not just to describe it but to verify it.

42. "Metaphysics" here is *ilāhiyyāt,* which means "divine philosophy."

43. Natural bodies, according to all Muslim Aristotelian philosophers, are bodies that have within themselves the principles of motion or stillness; this nature is part of the essence of the body and not accidental to it. See, for example, Abū al-Walīd Muḥammad Ibn Rushd, *Tafsīr mā ba'd al-Ṭabī'a,* 3 vols., ed. Maurice Bouyges (Beirut: Dār al-Shurūq, 1967), 838–39; Al-Kindī, *Rasā'il al-Kindī al-Falsafiyya* (Cairo, AH 1364), 165; and Ibn Sīnā, *Kitāb al-Najāt,* ed. M. Fakhry (Beirut: Dār al-Āfāq al-Jadīda, 1985), 135. Ibn Sīnā says that this concept of nature is not self-evident and needs demonstration, which is provided by the philosopher who deals with the First Philosophy. In contrast, Ibn Rushd argues that this concept of nature is self-evident, does not need an external proof, and can be proved within the science of natural philosophy. Ibn Rushd, *Tafsīr mā ba'd al-Ṭabī'a,* 508.

44. Al-Khafrī wrote *Al-Takmila fī Sharḥ al-Tadhkira,* a highly sophisticated commentary on Naṣīr al-Dīn al-Ṭūsī's *Tadhkira,* one of the classics of the eastern reform tradition. On Al-Khafrī see Saliba, "Sixteenth-Century Arabic Critique"; and Saliba, "Redeployment of Mathematics."

45. Saliba, "Redeployment of Mathematics," 119.

46. Saliba, "Redeployment of Mathematics," 120.

47. Nor did Al-Khafrī, who was a scholar of ḥadīth, seek to identify the model that would correspond best to a divinely ordained order, as Kepler did when he compared the models of Mars proposed by Ptolemy, Copernicus, and Tycho Brahe.

48. Jamil Ragep, "Ṭūsī and Copernicus: The Earth's Motion in Context," *Science in Context* 14, nos. 1–2 (2001): 145–63. Ragep's main argument is that Copernicus's discussion of the earth's motion falls within a mathematical tradition and employs arguments similar to those employed for a long time in Islam, and the arguments of both Copernicus and Muslim astronomers are

fundamentally different from the theological/natural philosophical arguments in medieval Europe (160).

49. Al-Qushjī, quoted in Ragep, "Ṭūsī and Copernicus," 156.

50. Ragep, "Ṭūsī and Copernicus," 156–57.

51. Saliba, "Redeployment of Mathematics," 120. Similarly, Al-Khafrī added several lunar models of his invention to those of Al-Ṭūsī and Al-Shīrāzī, but he did not suggest that only one was correct.

52. Ragep, "Ṭūsī and Copernicus," 154. In fact, neither Al-Bīrūnī before Al-Qushjī nor al-Khafrī after him needed or demanded such a conclusive proof.

53. *Naṣīr al-Dīn al-Ṭūsī's Memoir on Astronomy*, 1:107, 91.

54. *Naṣīr al-Dīn al-Ṭūsī's Memoir on Astronomy*, 1:38 ff.

55. Ragep, "Ṭūsī and Copernicus," 155–56.

56. Ragep, "Ṭūsī and Copernicus," 147–48.

57. See Robert Morrison, "*Mafhūm al-Uṣūl fī 'Ilm al-Hay'a fī al-Ḥadāra al-Islāmiyya*," in Bennacer El Bouazzati, ed., *Les éléments paradigmatiques thématiques et stylistiques dans la pensée scientifique* (Rabat: Publications of the Faculty of Letters and Human Sciences, 2004), 125–40, 137–39; and Morrison, "Quṭb al-Dīn al-Shīrāzī's Hypotheses for Celestial Motions," *Journal of Islamic Science* 20, nos. 1–2 (2004): 65–154; 21, nos. 1–3 (2005): 21–140. In the references below I quote the 2005 part of this essay.

58. Morrison convincingly argues that *uṣūl* translates better as "hypotheses" than as "principles." Morrison, "Quṭb al-Dīn al-Shīrāzī's Hypotheses," 24.

59. Morrison, "Quṭb al-Dīn al-Shīrāzī's Hypotheses."

60. Mūsā Ibn Maymūn, *Dalālat al-Ḥā'irīn*, ed. Husayn Atay (Ankara, n.d.), 345.

61. Abū Ishaq Nur al-Dīn al-Bitrūjī, *On the Principles of Astronomy* [*Kitāb fī al-Hay'a*], an edition of the Arabic and Hebrew versions, trans. Bernard Goldstein (New Haven: Yale University Press, 1977).

62. Al-Bitrūjī, *Kitāb fī al-Hay'a*, 49.

63. Ibn Rushd, *Tafsīr mā ba'd al-Ṭabī'a*, 1655–56.

64. Ibn Rushd, *Tafsīr mā ba'd al-Ṭabī'a*, 1664.

65. Ibn Rushd, *Tafsīr mā ba'd al-Ṭabī'a*, 285.

66. Ibn Rushd, *Talkhīs mā ba'd al-Ṭabī'a*, ed. Muḥammad 'Uthmān (Cairo, 1958), 131–32.

67. Ibn Rushd, *Talkhīs mā ba'd al-Ṭabī'a*, 133. See also Ibn Rushd, *Talkhīs al-Āthār al-'Ulwiyya*, ed. Jamāl al-Dīn al-'Alawī (Rabat: Dār al-Gharb

al-Islāmī, 1994), 202 ff. Ibn Rushd maintains that the shape of the heavenly bodies must be spherical because that shape suits their essence and nature.

68. In one respect, Copernican astronomy was also a counterrevolution, though much more creative than the Maghribī one. The justification for the new Copernican astronomy was not derived from observations, nor did Copernicus have at his disposal Kepler's and Newton's new physics. Instead, he drew on old natural philosophical principles (such as simplicity) to support his heliocentric model. This can be seen as a step back from mathematical experimentation, which is how he arrived at his models in the first place, to philosophical determinism. The assertion that the heliocentric models are the right models and not just possible, mathematically valid alternatives was a leap of faith on Copernicus's part, which he undertook in the name of old metaphysical principles that, for centuries before Copernicus, mathematical astronomers had learned to ignore.

69. Salem Yafut maintains that Ibn Rushd's and Al-Bitrūjī's critique of Ptolemaic astronomy laid the foundation for the Copernican reform. Nothing can be further from the truth: if there was any positive influence by the western Islamic astronomical tradition, it would be in the reaction against Ibn Rushd, and not in any acceptance of his arguments. Salem Yafut, *Naḥnu wal-'Ilm: Dirāsāt fī Tārīkh 'Ilm al-Falak bil-Gharb al-Islāmī* (Beirut, 1995).

70. See Ibn Rushd, *Talkhīṣ al-Athār al-'Ulwiyya*, ed. Jamāl al-Dīn al-'Alawī (Rabat: Dār al-Gharb al-Islāmī, 1994), 143–46. Ibn Rushd says that Aristotle does not mention these matters, not because he fails to be comprehensive, but because these details belong to another partial discipline. Ibn Rushd praises God, who privileged Aristotle with human perfection.

71. Ibn Rushd, *Tahāfut al-Tahāfut*, ed. Muḥammad al-'Uraybī (Beirut: Dār al-Fikr al-Lubnānī, 1993), 33.

72. Ibn Rushd, *Tahāfut al-Tahāfut*, 231. In reference to medicine, for example, Ibn Rushd maintains that health is achieved not through the application of the craft of medicine but by combining medicine and natural science, and that reforming medicine (just as he thought of reforming astronomy) can be accomplished by re-subjecting it to demonstrative proof (*al-qawl al-burhānī*). In other words, the principles derived from natural philosophy take precedence over those derived from within the craft of medicine. See Ibn Rushd, *Fī Ḥīlat al-Bar' 'alā Jālīnūs*, in *Rasā'il Ibn Rushd al-Ṭibbiyya*, ed. George Qanawātī and Sa'īd Zāyid (Cairo: Al-Hay'a al-Miṣriyya al-'Amma lil-Kitāb, 1987), 434 and passim; and Ibn Rushd, *Tahāfut al-Tahāfut*, 285.

73. Ibn Rushd composed three kinds of expositions of the works of

Aristotle. *Jawāmi‘* (often translated as "abridgements" or "epitomes" or "paraphrases"), *talākhīṣ* (middle summaries), and *shurūḥ* (grand summaries). The demonstrative dimension of Aristotelianism is outlined in jawāmi‘ and shurūḥ, whereas the talākhīṣ followed the same presentation found in the works of Aristotle. Jamāl al-Dīn al-‘Alawī, introduction to Abū al-Walīd Ibn Rushd, *Talkhīṣ al-Samā' wal-‘Ālam*, ed. Al-‘Alawī (Fes: Kulliyat al-Ādāb, 1984), especially 39–41. The medieval European fixation on rediscovering the "pure" Aristotle, and on purging Aristotelianism from the accretions introduced to it by commentators, especially from Averroësism, was itself a product of the very same Averroësism, which provided the most comprehensive vehicle for understanding the Aristotelian system in medieval Europe.

74. See, for example, Ibn Rushd, *Tahāfut al-Tahāfut*, 59–60, 114, 122, 224. This book was written in response to Al-Ghazālī's devastating critique of philosophy, *Tahāfut al-Falāsifa;* rather than taking issue with Al-Ghazālī, however, Ibn Rushd mostly argues that Al-Ghazālī's criticism applies to the erroneous concepts wrongly imposed on philosophy by Abū Naṣr al-Fārābī and Ibn Sīnā. In effect, the book is as much a critique of Al-Fārābī's and Ibn Sīnā's understanding of philosophy as it is of Al-Ghazālī's attack on it.

75. Ibn Rushd, *Tahāfut al-Tahāfut*, 142.

76. For a fuller discussion of this subject see Ahmad Dallal, "Women, Gender and Sexuality: Pre-Modern Scientific Discourses on Female Sexuality," in Suad Joseph, ed., *Encyclopedia of Women and Islamic Cultures* (Leiden: Brill, 2006), 3:401–7.

77. See Abū ‘Ali al-Ḥusayn b. ‘Abd Allāh Ibn Sīnā, *Al-Qānūn fī al-Ṭibb* (reprint of the Būlāq edition; Beirut: Dar Ṣādir, n.d.), Kitāb 3, Fann 21, Maqāla 1, Faṣl 1; and Ibn Rushd, *Al-Kulliyyāt fī al-Ṭibb, ma‘ Mu‘jam al-Muṣṭalaḥāt al-Ṭibbiyya al-‘Arabiyya*, ed. Muḥammad ‘Ābid al-Jābirī (Beirut: Markaz Dirasāt al-Waḥda al-‘Arabiyya, 1999), 157.

78. See Danielle Jacquart and Claude Thomasset, *Sexuality and Medicine in the Middle Ages*, trans. Matthew Adamson (Cambridge, UK: Polity Press, 1988), 54; and B. F. Musallam, *Sex and Society in Islam* (Cambridge: Cambridge University Press, 1983), 43–46.

79. Ibn Sīnā, *Al-Qānūn*, Kitāb 1, Fann 1, Ta‘līm 5, Faṣl 1; and Musallam, *Sex and Society in Islam*, 47–48.

80. Ibn Sīnā, *Qānūn*. Kitāb 1, Fann 3, Faṣl 1; and Kitāb 3, Fann 20.

81. Ibn Rushd, *Al-Kulliyyāt*, 187–90.

82. See Al-‘Urḍī, *Kitāb al-Hay'ah;* hereafter cited in the text. The introduction by George Saliba includes a comprehensive study of Al-‘Urḍī and his

position in the astronomical reform tradition. Many of the other studies by Saliba provide comparative analysis of the main figures in this astronomical tradition, including Al-'Urḍī.

83. See Al-'Urḍī, *Al-Fikr al-'Ilmi*, 95, 73–139, 116, 149, 156.

84. See, for example, the proof of the sphericity of the earth. Al-'Urḍī, *Kitāb al-Hay'ah*, 38–43.

85. See, for example, Al-'Urḍī, *Kitāb al-Hay'ah*, 64, 213, 215, 217, 106, 110, 121, 217.

86. See, for example, Al-'Urḍī, *Kitāb al-Hay'ah*, 110.

87. In one other similar instance, Al-'Urḍī talks of an impossibility according to natural science (*muḥāl min qibal al-'ilm al-ṭabī'ī*). Al-'Urḍī, *Kitāb al-Hay'ah*, 212. These references are noted here for their infrequency, whereas the references to uṣūl internal to the discipline are pervasive.

88. For an analysis of Al-'Urḍī's model see Saliba's introduction to Al-'Urḍī, *Kitāb al-Hya'ah*, 50–55.

89. The irony in this old notion of realism was that positing an identity between a mathematical model and reality was based on a largely metaphysical conception of reality.

90. Al-Bitrūjī, *Kitāb fī al-Hay'a*, 79–81.

91. The discussion so far is focused on the epistemological coherence of the astronomical tradition of research, but parallel, though not identical, trends can be traced in other sciences. The work of Rushdi Rashed provides an extremely rich source for tracing conceptual developments in the mathematical fields.

92. The main Arabic words used for classification are *taṣnīf* (categorization), *taqsīm* (division), and *tartīb* (ordering).

93. Modes of acquiring knowledge are, for example, memory (as in history), imagination (as in literature), and natural, human, or divine reason.

94. Ibn Sīnā, *Tis' Rasā'il* (Cairo: Hindiyya Press, 1908), 79–81. Similarly, Fārābī, defines *'aql* as a separate intellect which does not reside in matter. Abū Naṣr al-Fārābī, *Risāla fī al-'Aql*, ed. Maurice Boyges (Beirut: Al-Maktaba al-Kāthulikiyya, 1983), 21–27.

95. See, for example, Al-Fārābī, *Iḥṣā' al-'Ulūm*, ed. Uthmān Amīn (Cairo: Librairie Anglo-Égyptienne, 1968); Ibn Sīnā, *'Uyūn al-Ḥikma*, ed. 'Abd al-Rahmān Badawī (Beirut, 1980), 17; and Ibn Sīnā, *Manṭiq al-Mashriqiyyīn* (Beirut: Dār al-Ḥadātha, 1982).

96. Ibn Sīnā, *Al-Najāt*, ed. Majid Fakhry (Beirut: Dār al-Āfāq al-Jadīda, 1985), 135. As we have seen, Ibn Rushd carried this agenda to an extreme in his

attempt to restructure all the sciences on the basis of kulliyāt (universal principles). Earlier philosophers, such as Al-Fārābī and Ibn Sīnā, compromised, he thought. For a clear statement of his views see Ibn Rushd, *Tafsīr mā ba'd al-Ṭabī'a*, 298, 299, 309, 337, 339, 699, 701, 702, 714; for a critique of Ibn Sīnā see p. 1425.

97. See Muḥammad 'Ābid al-Jābirī, *Al-'Aql al-Akhlāqī al-'Arabī* (Casablanca: Al-Markaz al-Thaqāfī al-'Arabī, 2001), 111–13, who quotes the Mu'tazilī al-Qāḍī 'Abd al-Jabbār and the Ash'arī Fakhr al-Dīn al-Rāzī.

98. Al-Ghazālī, *Mi'yār al-'Ilm fī Fann al-Manṭiq*, ed. Sulaymān Dunyā (Cairo: Dār al-Ma'ārif, 1961), 33–41. Al-Ghazālī famously referred to the law as an external reason and to reason as an internal law (*al-shar' 'aql min khārij, wal 'aql shar' min dākhil*).

99. Ibn Taymiyya, *Nuṣūṣ*, ed. Aziz Azmeh (Beirut: Riad El-Rayyes Books, 2000), 229, 232–34.

100. Ibn Khaldūn, *Muqaddimat Ibn Khaldūn*, 3 vols., ed. 'Alī 'Abd al-Wāḥid Wāfī (Cairo: Nahḍat Miṣr, 2004), 3:968.

101. Ibn Taymiyya, *Dar' Ta'ārud al-'Aql wal-Naql*, ed. M. R. Salem (Cairo: Dār al-Kutub, 1971), 216.

102. Ibn Qayyim al-Jawziyya, *Kitāb al-Rūḥ*, ed. 'Ārif al-Ḥāj (Beirut, 1988), 364.

103. "Moreover, it is possible that the mind's operation relevant to the first intelligible things that are identical with individual occurrences is by means of imagined forms (*ṣuwar khayāliyya*) . . . in which case the judgment [of the mind] is certain and equivalent to perceptible things. . . . In this case we grant their claims." Ibn Khaldūn, *Muqaddima*, 3:1082–83.

104. Ibn Khaldūn, *Muqaddima*, 3:1082–83.

105. Of course, Ibn Khaldūn does not hesitate to pass moral judgments, but these do not relate to the possibility of knowledge, only to its effects. The possibility of knowledge, however, is determined on epistemological grounds.

106. Ibn Khaldūn, *Muqaddima*, 2:897.

107. Ibn Khaldūn, *Muqaddima*, 2:856; 3:917.

108. Ibn Khaldūn, *Muqaddima*, 3:928. This kind of reasoning is in addition to the three standard kinds of intellection: discriminating reason (*al-'aql tamyīzī*), which discerns the order of things either in nature or in conventional matters and roughly corresponds to the Aristotelian productive sciences; empirical/experimental reason (*al-'aql al-tajrībī*), the reasoning that provides

humans with the judgments and codes for dealing with fellow humans and corresponds to the Aristotelian practical sciences; and theoretical reasoning (*al-ʿaql al-naẓarī*), the reasoning that produces certain or possible knowledge of something beyond sense perception, that is not contingent on practice, and that corresponds to the Aristotelian theoretical sciences. Ibn Khaldūn, *Muqaddima*, 3:917.

109. Ibn Khaldūn, *Muqaddima*, 3:1080 f.

110. Ibn Khaldūn, *Muqaddima*, 1:342.

Chapter 3. Science and Religion

1. See Bennacer El Bouazzati, "The Continuum of Knowledge and Belief," *Bulletin of the Royal Institute for Inter-Faith Studies* 4, no. 1 (2002): 7–24.

2. Significantly, the word for "science" in Arabic is *ʿilm*, which technically means "knowledge"; there is no exclusive word for science that corresponds to the English word. Rather, there are words for specific sciences, such as *ʿilm al-hya'a* (astronomy), *ʿilm al-ḥadīth* (the science/discipline of ḥadīth), and *ʿilm al-dīn* (the science of religion).

3. The earliest interests in Islamic theology were about politics; the earliest theological schools predate the beginning of translation by several decades. Of course, Muslims could have been influenced by the structure of developed Christian theological arguments before having direct access to Greek philosophical works, especially dialectics. Also, it should be noted that the word "theology" does not completely correspond to the discipline in which theological issues were raised: *kalām*. The scope of kalām was considerably broader than than is implied in a narrow definition of theology, and often included discussion of issues that technically belong to the realm of science.

4. George Saliba, *Al-Fikr al-ʿIlmī al-ʿArabī; Nash'atuhu wa Taṭawwuruhu* (Balamand, Lebanon: Balamand University, 1998), 159–60. In several articles, Saliba delineates the various criticisms of astrology in Islam. See George Saliba, "The Cultural Context of Arabic Astronomy: Attacks on Astrology and the Rise of the *Hay'a* Tradition," *BRIIFS* 4, no. 1 (2002): 25–46; Saliba, "The Role of the Astrologer in Medieval Islamic Society," *Bulletin d'Études Orientales: Sciences occultes et Islam* 44, special issue coordinated by A. Regourd and P. Lory (1992): 45–68; Salib, "Astronomy and Astrology in Medieval Arabic Thought," in Roshdi Rashed and Joel Biard, eds., *Les doctrines de la science de l'antiquité à l'âge classique* (Louvain: Peeters, 1999), 131–64; Saliba, "The Ash'arites and the Science of the Stars," in Richard Hovannisian and Georges

Sabagh, eds., *Religion and Culture in Medieval Islam* (Cambridge: Cambridge University Press, 1999), 79–92. In this last essay Saliba notes that astrology was an integral part of the Aristotelian philosophical system; as such, the critique of astrology implicates Aristotelian natural philosophy. Saliba surveys the critiques of astrology by three theologians of the mainstream Ash'arī school, Abū Bakr al-Bāqillānī (d. 1013), Al-Ghazālī (d. 1111), and Sayf al-Dīn al-Āmidī (d. 1233). See also John Livingston, "Ibn Qayyim al-Jawziyyah: A Fourteenth Century Defense against Astrological Divination and Alchemical Transmutation," *Journal of the American Oriental Society* 91, no. 1 (1971): 96–103. Also see the chapter in the theological work by Abū Bakr al-Bāqillānī, *Kitāb al-Tamhīd*, ed. Richard McArthy (Beirut: Al-Maktaba al-Sharqiyya, 1957), 48–59, in which he criticizes astrology; and Yahya Michot, "Ibn Taymiyya on Astrology: Annotated Translation of Three Fatwas," *Journal of Islamic Studies* 11, no. 2 (2000): 147–208. Scholars have noted the very long list of distinguished classical Islamic writers who attacked astrology, including mathematicians, astronomers, philosophers, religious scholars, theologians, grammarians, and belletrists. See Michot, "Ibn Taymiyya on Astrology," 151.

5. David King, "The Astronomy of the Mamluks," *Isis* 74 (1983): 550. Also see King, "Mamluk Astronomy and the Institution of the Muwaqqit," in T. Philipp and U. Haarman, eds., *The Mamluks in Egyptian Politics and Society* (Cambridge: Cambridge University Press, 1998), 155.

6. Abū [al-]Rayḥān Muḥammad ibn Aḥmad al-Bīrūnī, *Taḥqīq mā lil-Hind min Maqūla Ma'qūla fī al-Aql am Mardhūla* (Beirut: 'Ālam al-Kitāb, n.d.), 219–21. For an English edition see *Alberuni's India*, 2 vols., trans. C. Edward Sachau (reprint, 1910; London: Trubner, 1887–88).

7. Muḥammad Ḥusayn al-Dhahabī, *Al-Tafsīr wa al-Mufassirūn*, 2 vols. (Cairo: Maktabat Wahba, 1985), especially 2:454–96. The following discussion of tafsīr is based on my article "Science and the Qur'ān," in *Encyclopaedia of the Qur'ān*, vol. 4 (Leiden: Brill, 2004), 540–58.

8. Muḥammad Iqbāl, *The Reconstruction of Religious thought in Islam* (Lahore: Institute of Islamic Culture, 1982), 30.

9. Al-Rāzī was a prolific and influential theologian of the Ash'arī school, which became the dominant theological school of Sunni Islam. Though mainly a religious scholar, Al-Rāzī was greatly influenced by philosophy and wrote, in addition to his tafsīr and major works in Islamic law, a commentary on Ibn Sīnā's philosophical work *Al-Shifā'*, as well as several other theological and philosophical treatises. On the life of Al-Rāzī see Frank Griffel, "On Fakhr al-Dīn al-Rāzī's Life and the Patronage He Received,"

Journal of Islamic Studies 18, no. 3 (2007): 313–44. On the thought of Al-Rāzī see Ayman Shihadeh, *The Theological Ethics of Fakhr al-Dīn al-Rāzī*, Islamic Philosophy, Theology and Science: Texts and Studies, vol. 64 (Leiden: Brill, 2006).

10. I use Ahmed Ali's translation of the Qur'ān (Princeton, NJ: Princeton University Press, 1984), with occasional changes of my own.

11. Fakhr al-Dīn al-Rāzī, *Al-Tafsīr al-Kabīr* (Cairo: Al-Maṭba'a al-Bahiyya al-Miṣriyya, 1934–62), 13–14:96 ff.; hereafter cited in the text. See also Abū Ḥayyān al-Andalusī, *Al-Nahr al-Mād min al-Baḥr al-Muḥīṭ*, ed. Būrān al-Ḍinnāwī and Hadyān al-Ḍinnāwī (Beirut: Dār al-Jinān, 1987), 1:809–11.

12. One of Al-Rāzī's books is entitled *Al-Mabāḥith al-Mashriqiyya fī 'Ilm al-Ilāhiyyāt wal-Ṭabī'iyyāt* (Eastern Studies in Metaphysics and Natural Science).

13. See, for example, Abū 'Abd Allāh Muḥammad b. Aḥmad al-Anṣārī al-Qurṭubī, *Al-Jāmi' li Aḥkām al-Qur'ān*, 3rd edition (Cairo: Dār al-Kitāb al-'Arabi lil-Tiba'a wal-Nashr, 1967), 2:191–202; Abū Ḥayyān, *Al-Nahr al-Mād min al-Baḥr al-Muḥīṭ*, 1:156 f.; Al-Rāzī, *Al-Tafsīr al-Kabīr*, 1–2:101 ff.; 8–10:137; 17–18:169.

14. See also Al-Qurtubi, *Al-Jāmi' li Aḥkām al-Qur'ān*, 7:230 ff.; 8: 38; Al-Rāzī, *Al-Tafsīr al-Kabīr*, 15–16:76; 17–18:37; Maḥmūd b. 'Umar al-Zamakhsharī, *Al-Kashshāf 'an Ḥaqā'iq Ghawāmid al-Tanzīl wa 'Uyūn al-Aqāwīl fī Wujūh al-Ta'wīl* (Cairo: Būlāq, 1864), 1:291, 354–55; Abū Ḥayyān, *Al-Nahr al-Mād min al-Baḥr al-Muḥīṭ*, 1.2:7; 2.1:49–50.

15. Al-Zamakhsharī, *Al-Kashshāf*, 1:43; Abū Ḥayyān, *Al-Nahr al-Mād min al-Baḥr al-Muḥīṭ*, 1:54.

16. See also Robert Morrison, "The Portrayal of Nature in a Medieval Qur'ān Commentary," *Studia Islamica* (2002): 20–22, for the different views of Al-Nīshābūri.

17. For example, Al-Rāzī, *Al-Tafsīr al-Kabīr*, 21–22:161–62.

18. For example, Al-Rāzī, *Al-Tafsīr al-Kabīr*, 21–22:163.

19. Al-Rāzī represents an important stage in the development of Muslim philosophical theology. On his impact see Ayman Shihadeh, "From al-Ghazālī to al-Rāzī: 6th/12th Century Developments in Muslim Philosophical Theology," *Journal of Arabic Sciences and Philosophy* 15, no. 1 (2005): 141–79; see both the study and the references therein. The present discussion of kalām draws on my essay "The Adjustment of Science," *BRIIFS* 4, no. 1 (2002): 97–108.

20. Alnoor Dahnani convincingly argues that physics in Islamic history

cannot be properly understood without the examination of works on kalām. Some of these works cannot be considered just, or even primarily, works of theology, but legitimate works of physics worthy of study by historians of science. See Alnoor Dhanani, *The Physical Theory of Kalām: Atoms, Space, and Void in Baṣrian Mu'tazilī Cosmology* (Leiden: Brill, 1994); and Dhanani, "Problems in Eleventh Century *Kalām* Physics," *BRIIFS* 4, no. 1 (2002): 75–96.

21. In contrast, later mathematical works by Ibn al-Bannā' and Bahā' al-Dīn al-'Āmilī (d. 1621), among others, abound with metaphysical and mystical interpretations. See Ibn al-Bannā' al-Marākushī, *Raf' al-Ḥijāb 'an Wujūh A'māl al-Ḥisāb*, ed. Muḥammad Aballāgh (Fes: Kulliyat al-Ādāb, 1994); and Bahā' al-Dīn al-'Āmilī, *Al-A'māl al-Riyādiyya al-Kāmila*, ed. Jalāl Shawqī (Beirut: Dār al-Shurūq, 1981).

22. By his own testimony, Ibn Rushd's education in astronomy was negligible, and it shows. See Ibn Rushd, *Tafsīr ma ba'd al-Ṭabī'a*.

23. See, for example, Abū al-Ḥasan al-Ash'arī (d. 936), *Maqālāt al-Islāmiyyīn wa Ikhtilāf al-Muṣallīn*, ed. H. Ritter (Istanbul: Maṭba'at al-Dawla, 1929–30).

24. Ibn Taymiyya (d. 1328) was a mathematician from a family of distinguished mathematicians. Distinguished religious scholars who wrote on theology as well as advanced astronomy include Niẓām al-Dīn al-Nīshābūrī (d. c. 1330), Ṣadr al-Sharī'a al-Bukhārī (d. 1347), Al-Sharīf al-Jurjānī (d. 1413), Fath Allāh al-Shirwānī (c. 1440), and Shams al-Dīn al-Khafrī (d. 1550). See, for example, Saliba, *Al-Fikr al-'Ilmī*, 161 and passim.

25. For this argument see A. I. Sabra, "Science and Philosophy in Medieval Islamic Theology: The Evidence of the Fourteenth Century," *Zeitschrift für Geschichte der Arabisch-Islamischen Wissenschaften* 9 (1994): 1–42. For a critique of this view see Dallal, "Adjustment of Science." Sabra's essay draws mainly on the work of 'Aḍuḍ al-Dīn al-Ijī (d. 1355), *Kitāb al-Mawāqif fī 'ilm al-Kalām* (Beirut: 'Ālam al-Kitāb, n.d.); and on the commentary by Al-Sharīf al-Jurjānī on Al-Ijī's *Kitāb al-Mawāqif*. In the following analysis I will focus on a rereading of Al-Ijī's work.

26. Sabra, "Science and Philosophy in Medieval Islamic Theology," 9.

27. Sabra, "Science and Philosophy in Medieval Islamic Theology," 10; Sabra's emphasis.

28. Sabra, "Science and Philosophy in Medieval Islamic Theology," 19, 22.

29. Sabra, "Science and Philosophy in Medieval Islamic Theology," 24.

30. Al-Ijī, *Kitāb al-Mawāqif*, 4.

31. *Fa inna al-khasma, wa-in khaṭṭa'nāhu, la nukhrijahu min 'ulamā' al-kalām.*

32. Al-Ijī, *Kitāb al-Mawāqif,* 8.

33. See, for example, Al-Ijī, *Kitāb al-Mawāqif,* 8.

34. Al-Ijī, *Kitāb al-Mawāqif,* 8–9.

35. See, for example, Al-Ijī, *Kitāb al-Mawāqif,* 200 and passim.

36. Al-Ijī, *Kitāb al-Mawāqif,* 207.

37. See, for example, Al-Ijī, *Kitāb al-Mawāqif,* 203–4, 213.

38. Al-Ijī, *Kitāb al-Mawāqif,* 215.

39. Al-Ijī, *Kitāb al-Mawāqif,* 254–56.

40. *Inna min al-furūd ma yaḥkum al-'aql bi-jawāzihi, kal-furūd al-handasiyya, wa laysa li aḥad an yamna'ahu illā mukābaratan.*

41. Reference here is to Ṣadr al-Sharī'a al-Bukhārī, *Kitāb Ta'dīl al-'Ulūm,* Ms. B, India Office (London), Arab Ms. Loth 532; hereafter cited in the text. For more on Ṣadr al-Sharī'a, and for a detailed study of his astronomical work, see Ahmad Dallal, *An Islamic Response to Greek Astronomy: Kitāb Ta'dīl Hay'at al-Aflāk of Ṣadr al-Sharī'a* (Leiden: Brill, 1995).

42. As Sabra suggests; see, for example, Sabra, "Science and Philosophy in Medieval Islamic Theology," 32.

43. See Abū Ḥāmid al-Ghazālī, *The Incoherence of the Philosophers / Tahāfut al-Falāsifa, a Parallel English-Arabic Text,* ed. and trans. Michael Marmura (Provo, UT: Brigham Young University Press, 1997).

44. Other relevant works by Al-Ghazālī include *Al-Iqtiṣād fī al-I'tqād,* ed. I. A. Cubukcu and H. Atay (Ankara: Nur Matbaasi, 1962); *Mi'yār al-'Ilm fī al-Manṭiq,* ed. Aḥmad Shams al-Dīn (Beirut: Dār al-Kutub al-'Ilmiyya, 1990); *Miḥakk al-Naẓar fī al-Manṭiq,* ed. M. al-Na'sānī (Beirut: Dār al-Nahḍa, 1966); *Al-Qusṭās al-Mustaqīm,* in *Majmū'at Rasā'il al-Imām al-Ghazālī,* vol. 3 (Beirut: Dār al-Kutub al-'Ilmiyya, 1986); *Maqāṣid al-Falāsifa,* ed. Sulaymān Dunyā (Cairo: Dār al-Ma'ārif, 1961); and *Al-Mustaṣfā min 'Ilm al-Uṣūl,* 2 vols. (Būlāq: al-Maṭba'a al-Amīriyya, AH 1322). For an excellent recent summary of scholarship on Al-Ghazālī's life and work see Frank Griffel, "Al-Ghazali," in *Stanford Encyclopedia of Philosophy,* 1–28, http://plato.stanford.edu/al-ghazali/ (first published Aug. 14, 2007). For important and differing assessments of Al-Ghazālī's attitude toward the sciences see, for example, Michael Marmura, "Ghazālī and Demonstrative Science," *Journal of the History of Philosophy* 3 (1965): 183–204; Marmura, "Ghazālī's Attitude to the Secular Sciences and Logic," in G. Hourani, ed., *Essays on Islamic Philosophy and Science* (Albany: State University of New York Press, 1975), 185–215; Marmura, "Al-

Ghazālī's Second Causal Theory in the 17th Discussion of His *Tahāfut*," in P. Morewdge, ed., *Islamic Philosophy and Mysticism* (Delmar, NY: Caravan Books, 1981), 85–112; Marmura, "Ghazālian Causes and Intermediaries," *Journal of the American Oriental Society* 115 (1995): 89–100; Richard Frank, *Creation and the Cosmic System: Al-Ghazālī and Avicenna* (Heidelberg: C. Winter, 1992); Frank, *Al-Ghazālī and the Ash'arite School* (Durham, NC: Duke University Press, 1994); T. Kukkonen, "Possible Worlds in *Tahāfut al-Falāsifa:* Al-Ghazali on Creation and Contingency," *Journal of the History of Philosophy* 38 (2000): 479–502; and B. D. Dutton, "Al-Ghazālī on Possibility and the Critique of Causality," *Medieval Philosophy and Theology* 10 (2001): 23–46. For other important studies of Al-Ghazālī's thought see E. Ormsby, *Theodicy in Islamic Thought: The Dispute over al-Ghazālī's "Best of Possible Worlds"* (Princeton, NJ: Princeton University Press, 1984); and Ebrahim Moosa, *Ghazālī and the Poetics of Imagination* (Chapel Hill: University of North Carolina Press, 2005).

45. The edition consulted here is Al-Imam al-Ghazālī, *Tahāfut al-Falāsifa*, ed. Sulayman Dunya (Cairo: Dār al-Ma'ārif, 1958), 72; hereafter cited in the text. Al-Ghazālī says that he uses the views of ancient philosophers as reported on the authority of Al-Fārābī and Ibn Sīnā (76).

46. One example of the second kind is the explanation of a lunar eclipse.

47. For the doctrine of atomism and other theological doctrines see Shlomo Pines, *Beitrage zur islamischen Atomenlehre* (Berlin: Grafenhainichen, 1936); and Harry Wolfson, *The Philosophy of Kalām* (Cambridge: Harvard University Press, 1976).

48. As Frank Griffel puts it, "An occasionalist universe will always remain indistinguishable from one governed by secondary causality." See Griffel, "Al-Ghazali," 23. On causation in the works of Al-Ghazālī see L. E. Goodman, "Did al-Ghazālī Deny Causality?" *Studia Islamica* 47 (1978): 83–120; Michael Marmura, "Ghazālī and Ash'arism Revisited," *Journal of Arabic Science and Philosophy* 12, no. 1 (2002): 91–110; Marmura, "Ghazālī and Demonstrative Science," 183–204; Marmura, "Ghazālī's Attitude to the Secular Sciences and Logic," 100–111; also see George Makdisi, "Ash'arī and the Ash'arites in Islamic Religious Thought," *Studia Islamica* 17 (1962): 27–80; 18 (1963): 19–39.

49. Ibn Khaldūn, *The Muqaddimah*, trans. Franz Rosenthal, abridged by N. J. Dawood (Princeton, NJ: Princeton University Press, 1967), 143.

50. In contrast, many historians maintain that the Ikhwān al-Ṣafā (10th c.) inaugurated in their book, *Rasā'il Ikhwān al-Ṣafā*, an Islamic cosmological

tradition that continued after their demise. Ikhwān al-Ṣafā, however, was a secret cult that had little impact on, and was criticized in the Islamic intellectual tradition by, religious scholars and Aristotelian philosophers and scientists. As Abū Ḥayyān al-Tawḥīdī (d. 1023) says in his *Al-Imtā' wal-Mu'ānasa*, ed. Aḥmad Amīn (Cairo, 1939–40), the group's objective was to merge Greek philosophy and Arabic religious law (*sharī'a*); the cosmological dimension of their work was therefore a truncated version of Greek cosmology imposed on Islamic law. See, for example, Abbas Hamdani, "Abū Ḥayyān al-Tawḥidī and the Brethren of Purity," *International Journal of Middle East Studies* 9, no. 3 (1978): 345–53. To be sure, an argument can be made about the emergence of specifically Islamic mystical cosmologies in the late Middle Ages, but these had followers and representatives and have nothing to do with Ikhwān al-Ṣafā. On the other hand, the alternative cosmologies discussed here emerged in the Islamic cultural space as "scientific" cosmologies that competed with Greek cosmology and were not advocated in the name of religion.

51. See the sections on the sciences of magic, talismans, secret letters, alchemy, astrology, and the refutation of philosophy in Ibn Khaldūn, *Al-Muqaddima* (Beirut: Dār al-Kitāb al-Lubnānī, 1967), 924, 928–30, 938. Ibn Khaldūn, like many astronomers and scientists before him, disengages astronomy (or the Arabic tradition of hay'a) from astrology and argues, on epistemological grounds, that astrology is not a legitimate form of knowledge, although it is in the Greek philosophical tradition.

52. Ibn Khaldūn, *Al-Muqaddima*, 997, 843–44.

53. Ibn Khaldūn, *Al-Muqaddima*, 1001.

54. Ibn Khaldūn, *Al-Muqaddima*, 908–15.

55. Ibn Khaldūn, *Al-Muqaddima*, 274–75.

56. This neutral separation is not to be confused with Ibn Rushd's separation, which presupposes a definite hierarchy of knowledge, with philosophy at the top. Interestingly, Ibn Rushd poses the same question that the ḥadīth scholar Ibn al-Ṣalāḥ (d. 1245) asked, and that was prominently noted by Ignaz Goldziher: whether it is religiously lawful to study the philosophical sciences. Whereas Ibn al-Ṣalāḥ concludes that this study is prohibited, Ibn Rushd argues that it is obligatory or at least recommended. See Ibn Rushd, *On the Harmony of Religion and Philosophy* [*Kitāb Faṣl al-Maqāl*], trans. George Hourani (London: Luzac, 1976); and Ibn al-Ṣalāḥ, *Fatāwā* (Cairo, n.d.). The developments described in the first three chapters obviated the need for asking this question altogether.

57. Ibn Khaldūn, *The Muqaddimah*, trans. Rosenthal, 371.

Chapter 4. In the Shadow of Modernity

1. So far I have deliberately avoided the common and understandable focus on precedence in scientific discoveries. The many historians of Islamic science who focus on precedents run the risk of slipping into a form of chauvinistic nationalism. But the main problem is that precedents gain their meaning in their own social, cultural, and epistemological contexts, as part of the larger system of knowledge in which they are embedded. I have also deliberately avoided approaching the subject of Islamic science by focusing on the influence of Islamic science in Europe. That influence was, no doubt, crucial, but my aim in this book is to study Islamic science in its own environment and not based on a teleology that sees values in this science only to the extent that it influenced European science.

2. For such problematic approaches to the history of Islamic science see Toby E. Huff, *The Rise of Early Modern Science: Islam, China and the West* (New York: Cambridge University Press, 1993), passim. General histories of science contain frequent, matter-of-fact references to the decline of Islamic science; see, for example, Edward Grant, *A History of Natural Philosophy from the Ancient World to the Nineteenth Century* (Cambridge: Cambridge University Press, 2007), 92–93; Grant, *Physical Science in the Middle Ages* (Cambridge: Cambridge University Press, 2007); and David Lindberg, *The Beginnings of Western Science* (Chicago: Chicago University Press, 1992), 161–82. Historians' disproportionate interest in the decline of the Islamic sciences is reflected in a recent surge in articles on the relationship between science and Islam. In an essay in the *Times Literary Supplement*, Steven Weinberg asserts that after Al-Ghazālī, "there was no more science worth mentioning in Islamic countries." Steven Weinberg, "A Deadly Certitude on God, Christianity and Islam," *Times Literary Supplement*, January 17, 2007. In a response to Weinberg, Jamil Ragep succinctly sums up the problems in Weinberg's assertion: "Thus does Weinberg dismiss three or four generations of scholarship over the past hundred years that has brought to light the work of scores of Islamic scientists between the twelfth and eighteenth centuries who, among other things, proposed the idea of pulmonary circulation, built the first large-scale astronomical observatories, conceived trigonometry as a separate discipline, constructed new calculating devices and maps of astonishing accuracy and sophistication, allowed for the possibility of a moving Earth, developed the mathematical and conceptual tools that were essential for the Copernican revolution, and made science and mathematics a part of the school (madrasa)

curriculum. Given the considerable literature now available on these subjects, it is difficult to understand why Weinberg prefers ideologically-based opinion to solid historical research." Jamil Ragep, "Response to Weinberg," *Times Literary Supplement*, January 24, 2007. In his response to Ragep, Weinberg insists on his earlier arguments, although they are based on flagrant factual errors. Weinberg, "Response to Ragep," *Times Literary Supplement*, January 31, 2007. For example, without bothering to consult the considerable literature on the invention of trigonometry as a separate discipline by the thirteenth-century scholar Naṣīr al-Dīn al-Ṭūsī, Weinberg asserts that trigonometry was the work of Al-Battānī (d. 918) and Al-Bīrūnī (973–c. 1048). In fact, Weinberg seems to base all his knowledge about Islamic astronomy on the very outdated scholarship on Al-Battānī, whom he seems to think was the primary astronomer in the Islamic tradition. Thus, Weinberg absurdly insists that Copernicus chiefly relied on Al-Battānī's work, despite the large literature now made available both by Copernicus scholars (Noel Swerdlow and Otto Neugebauer) and by historians of Arabic astronomy (especially George Saliba and Jamil Ragep) that shows the indebtedness of Copernicus to thirteenth- and fourteenth-century Muslim astronomers. Weinberg also repeats clichés about the work of scientists after Al-Ghazālī, saying that they "found no place in Islamic society" and that the work of Ibn al-Nafīs (13th c.) on pulmonary circulation "had no effect in the Islamic world, perhaps because for religious reasons he did not demonstrate its truth by the dissection of animals." Had Weinberg bothered to consult any contemporary sources on Islamic science or relied on actual historical evidence, he would have refrained from making such assertions or would at least have tried to nuance his arguments. In fact, some of Weinberg's assertions can be overturned by simply relying on a rudimentary knowledge of science. Weinberg says, for example, that large-scale observatories were used for predicting prayer times and the Muslim lunar months, although neither large nor small observatories are needed to make the necessary calculations.

On the opposition between Islam and science see also Todd Pitock, "Science and Islam in Conflict," *Discover Magazine*, June 21, 2007; and Robert Irwin, "Islamic Science and the Long Siesta: Did Scientific Progress in the Islamic World Really Grind to a Halt after the Twelfth Century?" *Times Literary Supplement*, January 23, 2008.

3. The most common assertion is that the sciences declined because they were always marginal to the larger religious culture and in conflict with it; this is the theme of Goldziher's influential article and many studies since.

I. Goldziher, "The Attitude of Islamic Orthodoxy towards the Ancient Sciences," in Merlin Swartz, ed., *Studies in Islam* (New York: Oxford University Press, 1981), 185–215. Another theory is that appropriating the sciences and employing them in the service of religion inhibited the pursuit of science as an independent inquiry and thus curtailed their free development.

4. As I have argued here, the epistemological separation of science and religion produced a relatively autonomous scientific culture. Because it was fairly autonomous, the eventual decline of science cannot be explained in terms of its alleged conflict with religion, nor in terms of a restrictive appropriation, or deadly embrace, of science by religion.

5. Instead of assuming that the sciences started their decline around the time of Al-Ghazālī (d. 1111) because of the conflict between religion and science (which Al-Ghazālī supposedly articulated in his works), the historical record shows that scientific culture was vibrant until at least the sixteenth century. Recognizing a definite advance in the sciences for several centuries after Al-Ghazālī rules out the attribution of decline to Al-Ghazālī and his brand of hostility toward philosophy. Asserting the early decline of Islamic science makes it irrelevant for the study of the rise of modern science, thereby allowing historians of modern science to exclude Islamic science from their accounts. Accepting the continued vibrancy of Islamic scientific culture underscores the need to understand modern science as a product of a continuous scientific tradition.

6. Al-Maqarrī, *Nafḥ al-Ṭīb bi Akhbār al-Andalus al-Raṭīb* (Beirut, 1968), 4:817. In an earlier account, Ṣā'id al-Andalusī (d. 1070) says that endemic frontier wars exhausted the collective energies of Andalusians and caused a temporary recession in scientific activity, but he notes that scientific activity resumed after the restoration of order. Ṣā'id al-Andalusī, *Ṭabaqāt al-Umam*, ed. Hayāt Bū 'Alwān (Beirut: Dār al-Ṭalī'a, 1985), 165.

7. Ibn Khaldūn, *Muqaddimat Ibn Khaldūn*, 3 vols., ed. 'Alī 'Abd al-Wāḥid Wāfī (Cairo: Nahḍat Miṣr, 2004), 1:52–53. Ibn al-Wazzān (Leo Africanus, 16th c.) talks about the palaces of Marrakech, which were used as chicken and pigeon farms. Ibn al-Wazzān, *Waṣf Ifrīqiyā*, trans. M. Ḥijjī and M. Al-Akhḍar (Rabat, 1980), 1:106.

8. In Ibn Khaldūn's assessment, the sciences declined because scientists "got into the habit of [abstract] intellectual reflection and searched for meanings while dissociating these meanings from perceptibles (*intizā'ahā min al-maḥsūsāt*) and abstracting them in the mind as general universal principles (*tajrīdahā fī al-dhihn umūran kulliyatan 'āmmatan*) . . . so all of their judgments and reflections are mental (*fī al-dhihn*) and do not correspond with what is

perceived (*la taṣīr ilā al-muṭābaqa*)." In the rest of the matters the scientists consider, they are used to "mental [speculations] and intellectual reflections and know nothing else aside from these." This assessment was equally applicable to the legal and the rational sciences. Ibn Khaldūn, *Al-Muqaddima* (Beirut: Dār al-Kitāb al-Lubnānī, 1967), 1045–46.

9. Ibn al-Bannā' al-Marākishī, *Raf' al-Ḥijāb 'an Wujūh A'māl al-Ḥisāb*, ed. Muḥammad Aballāgh (Fes: Kuliyyat al-Ādāb, 1994). For similar examples see Bahā' al-Dīn al-'Āmilī (d. 1621), *Al-A'māl al-Riyādiyya al-Kāmila*, ed. Jalāl Shawqī (Beirut: Dār al-Shurūq, 1981).

10. For a different view see Salem Yafut, who provides a positive evaluation of the role of astrology and metaphysics in science: Salem Yafut, "*Madkhal li-Qirā'at 'al-Madkhal ilā Ṣinā'at Aḥkām al-Nujūm li-Ibn Bannā al-Marākishī*," in Bennacer El Bouazzati, ed., *Science et pensée scientifique en Occident Musulman au Moyen Age* (Rabat: Faculty of Letters, 2001).

11. For accounts of eighteenth-century educational reforms in the Ottoman empire, including the introduction of numerous scientific subjects to the school curriculum, see Ekmeleddin Ihsanoglu, "The Introduction of Western Science to the Ottoman World: A Case Study of Modern Astronomy (1660–1860)," in John Brooke and Ekmeleddin Ihsanoglu, eds., *Religious Values and the Rise of Science in Europe* (Istanbul: Research Centre for Islamic History, Art and Culture, 2006), 185–228. Also see Ihsanoglu, *Science, Technology and Learning in the Ottoman Empire: Influence, Local Institutions, and the Transfer of Knowledge* (Aldershot, UK: Ashgate, 2004). This reform effort involved importing European teachers and was focused primarily on technology; it is not clear, therefore, what effect it had on the development of theoretical science. The Ottoman *Tanẓīmāt* reform was paralleled by efforts at strengthening the state in Iran, Egypt, and Morocco. The need to modernize the army was an incentive for these reforms, but increasing the reach of the state and its ability to support the military reforms required the development of new fiscal systems and efficient bureaucracies.

12. Based on the 1840 treaty signed in London by the United Kingdom, Austria, Prussia, Russia, and the Ottoman empire.

13. See Khaled Fahmy, "The Era of Muhammad 'Ali Pasha," in M. W. Daly, ed., *The Cambridge History of Egypt*, vol. 2: *Modern Egypt from 1517 to the End of the Twentieth Century* (Cambridge: Cambridge University Press, 1998), 139–78. Also see Afaf Lutfi al-Sayyid Marsot, *Egypt in the Reign of Muhammad Ali* (Cambridge: Cambridge University Press, 1984). In Indonesia the Dutch closed Islamic institutions and banned Muslims from universities until 1952.

See Adrian Vickers, *A History of Modern Indonesia* (Cambridge: Cambridge University Press, 2005).

14. See, for example, the Economic and Social Commission for Western Asia (ESCWA) 2003 report, *New Indicators for Science, Technology and Innovation in the Knowledge-Based Society* (New York: United Nations, 2003), 16, 28, 66, and passim; and the section on science in United Nations Development Programme (UNDP), *Arab Human Development Report, 2002* (New York: UNDP, 2002), 65–83. Few Muslim governments collect data on the status of science and innovation. The Organization of the Islamic Conference (OIC) database on its fifty-seven member countries has no data on research.

15. The main exception to this general rule is Turkey, which has a significantly better performance in some of the science and technology indicators.

16. See, for example, ESCWA, *New Indicators,* 21 f.: "much of the production and services technologies have been acquired through technology transfer, both in the public and private sectors."

17. Between 1990 and 1996, the number of universities in the Arab region increased by 50 percent (from 117 to 175), and the number of private universities increased sevenfold. See ESCWA, *New Indicators,* 16–17; and A. Zahlan, *Al-'Arab wa Taḥaddiyāt al-'Ilm wa-l-Taqāna. Taqaddum min dūn Taghayyur* (Beirut: Markaz Dirāsāt al-Wiḥda al-'Arabiyya, 1999), 18. For information on education and the low number of researchers as a percentage of the population see the World Bank, *World Development Indicators* (WDI), devdata.worldbank.org/wdi2006/contents/index2.htm. The highest recorded expansion of higher education in Organization of the Islamic Conference (OIC) countries happened in Pakistan and Iran: Iran's university student population increased from 100,000 in 1979 to 2,000,000 today. Seventy percent of science students are women, and the student-lecturer ratio decreased from 36:1 in 1989 to 18:1 in 2006. Pakistan's university student population increased from 276,000 in 2001 to 423,000 in 2004. Nonetheless, in Iran and Pakistan, as in other OIC countries, the economies fail to absorb college graduates.

18. See, for example, A. Zahlan, *Al-'Ilm wa-l-Siyāsa al-'Ilmiyya fī al-Waṭan al-'Arabī,* 5th edition (Beirut: Markaz Dirāsāt al-Wiḥda al-'Arabiyya, 1990); Zahlan, *Science and Science Policy in the Arab World* (London: Croom Helm, 1980); Zahlan, *Al-'Arab wa Taḥaddiyāt al-'Ilm wa-l-Taqāna;* and Ziauddin Sardar, *Science and Technology in the Middle East* (Essex, UK: Longman, 1982), a slightly outdated but useful guide to issues, organizations, institu-

tions, and country profiles. Also see the section on science in UNDP, *Arab Human Development Report*, 65–83.

19. These indicators include the number of research institutions, scientific publications and patents, and the status of higher education.

20. Again, the two exceptions in the Muslim world are Turkey and Malaysia, whose spending is comparable to that of other moderately wealthy nations. Malaysia in particular is technologically advanced, invests heavily in the electronics industry, and is a high-tech exporter. Under military rulers, the spending on science and technology research in Pakistan has increased significantly. Most of this funding, however, is devoted to military technologies. Similarly, Iran invests in military research, but many other areas of science, such as stem cell research, are also well funded. Among the Organization of the Islamic Conference countries, Saudi Arabia and Kuwait, on the other hand, are among the lowest in research and development spending as a percentage of the gross domestic product. For information on research and development spending as a percentage of GDP see the Statistical, Economic and Social Research and Training Center for Islamic Countries (SESRTCIC) Web site, www.sesrtcic.org; and reports at the UNESCO Statistics Division Web site, stats.uis.unesco.org.

21. UNDP, *Arab Human Development Report*, 65–66. Also see Zahlan, *Al-'Arab wa Taḥaddiyāt al-'Ilm wa-l-Taqāna*, 18, 28. Zahlan notes that in a population of about 50,000 scientists holding a Ph.D., the total number of publications is about 6,000 scientific papers per year (95). Also see National Science Board, *Science and Engineering Indicators, 2006* (Arlington, VA: National Science Foundation, 2006), www.nsf.gov/statistics/seind06. The average production of scientific articles in 2003 for all countries was 137; the average for Organization of the Islamic Conference (OIC) countries was 13. Once again, the exceptions to this general pattern were Turkey (up from 500 articles in 1988 to more than 6,000 per year in 2003) and Iran (up from 100 articles per year a decade ago to 2,000 today).

22. See the description of the mission of ICPSR at http://www.icpsr.org.ma/?Page=icpsr. An interesting development is the appointment of Ekmeleddin Ihsanoglu, a historian of Ottoman science and the former president of the International Union for the History and Philosophy of Science, as the secretary general of the Organization of the Islamic Conference.

23. Ernest Renan, "L'islamisme et la science," *Journal des Débats* (Mar. 29, 1883).

24. Renan's hatred of the subject of his study is notable, and such hatred is not uncommon among Orientalists. But examining the reasons behind this hatred is beyond the scope of this study.

25. Nikki Keddie, *An Islamic Response to Imperialism: Political and Religious Writings of Sayyid Jamāl al-Dīn "al-Afghānī"* (Berkeley: University of California Press, 1983). For the response to Renan see especially 182–84, 187; also see 102–3, 104–7.

26. Ḥasan al-Bannā, "Towards the Light," in Al-Bannā, *Five Tracts of Ḥasan al-Bannā' (1906–1949): A Selection from the Majmū'at Rasā'il al-Imām al-Shahīd Ḥasan al-Bannā'*, trans. Charles Wendell (Berkeley: University of California Press, 1978).

27. For example, Sayyid Quṭb, *Social Justice in Islam*, trans. John B. Hardie; rev. Hamid Algar (Oneonta, NY: Islamic Publications International, 2000), 27–28, and passim. Elsewhere Quṭb maintains that "in fact it was Islam, by virtue of its realistic system that initiated the experimental method, which was passed to Europe from the Universities of Andalusia, and on which, Roger Bacon and Francis Bacon, falsely alleged to be the fathers of this school, founded their methods." Sayyid Quṭb, *Al-Mustaqbal li-hādhā al-Dīn* (Cairo: Dār al-Shurūq, 1991), 85.

28. In some of his writings, Sayyid Quṭb maintains that Muslims cannot live in isolation from science, but that the sciences are a mixed blessing and can cause harm; Quṭb also recognizes the philosophical (and epistemological) foundations of modern science. See Quṭb, *Social Justice in Islam*, 286–89. However, in his most influential work, *Milestones*, he reverts to a universal notion of science as having no specific cultural traits and no specific attachment to the West. Sayyid Quṭb, *Ma'ālim fī al-Ṭarīq* (Cairo: Dār al-Shruūq, n.d.), 6–7.

29. Quṭb, *Ma'ālim fī al-Ṭarīq*. 7. In this regard, Islamists and secular nationalists held almost identical views. The United Arab Republic Charter, drafted under Jamal 'Abd al-Nasser, asserts that "revolutionary action should be scientific. . . . Science is the true weapon of revolutionary will. Here emerges the great role to be undertaken by the universities and educational centers on various levels. . . . Science is the weapon with which revolutionary triumph can be achieved." *The United Arab Republic Charter* (1958), 88–89.

30. Some of these works discussing Copernican astronomy include reference to the kalām work of Al-Ījī, discussed in the previous chapter, but not to the technical works on astronomy.

31. On the relationship between the Islamic reform tradition of theo-

retical astronomy and Copernicus see Noel Swerdlow and Otto Neugebauer, *Mathematical Astronomy in Copernicus's De Revolutionibus* (New York: Springer-Verlag, 1984); George Saliba, "Arabic Astronomy and Copernicus," *Zeitschrift für Geschichte der Arabisch-Islamischen Wissenschaften* 1 (1985): 73–87; and Jamil Ragep, "Copernicus and His Islamic Predecessors: Some Historical Remarks," *History of Science* 45, no. 1 (2007): 65–81. These scholars analyze in detail the Islamic foundations of Copernican mathematical astronomy.

32. *Al-Muqtaṭaf*, part 1 (1876): 171–74; George Saliba, *Al-Fikr al-ʻIlmī al-ʻArabī. Nash'atuhu wa Taṭawwruhu* (Balamand, Lebanon: Balamand University, 1998), 138–39; and George Saliba, "Copernican Astronomy in the Arab East: Theories of the Earth's Motion in the Nineteenth Century," in Ekmeleddin Ihsanoglu, ed., *Transfer of Modern Science and Technology to the Ottoman World* (Istanbul: Research Centre for Islamic History, Art and Culture, 1992).

33. ʻAbd Allah Fikrī, letter in *Al-Muqtaṭaf*, part 1 (1876): 219.

34. For an excellent discussion of the reception of both the new astronomy and Darwinism in Beirut see Marwa Elshakry, "The Gospel of Science and American Evangelism in Late Ottoman Beirut," *Past and Present* 196, no. 1 (2007): 173–214.

35. Shiblī Shumayyil, *Falsafat al-Nushū' wal-Irtiqā'* (Cairo: Maṭbaʻat al-Muqtaṭaf, 1910). This was essentially a translation of Ludwig Buchner's commentary on Darwin.

36. Incidentally, Shumayyil was also adamant in his belief in the inferiority of women and that a man's brain is bigger than a woman's.

37. In 1880, Father Jirgis Faraj published a book refuting the principle of natural selection; and in 1886 Ibrāhīm Hūrānī rejected evolution on the grounds that it is based on chance and guesswork. See Adel Ziadat, *Western Science in the Arab World: The Impact of Darwinism, 1860–1930* (London: St. Martin's Press, 1986).

38. Shumayyil also maintained that "Egypt under British occupation developed its irrigation system and agriculture; the life of the peasants became richer . . . and its finances better organized . . . to the extent that it gained the confidence of the world and freedom prevailed." Quoted in Najm Bezirgan, "The Comparative Reception of Darwinism: The Islamic World," in Thomas Glick, ed., *The Comparative Reception of Darwinism* (Chicago: University of Chicago Press, 1988), 382.

39. Jamāl al-Dīn al-Afghānī and Muḥammad ʻAbdu, *Al-ʻUrwa al-Wuthqā* (Beirut: Dār al-Kitāb al-ʻArabī, 1980), 414.

40. Al-Afghānī and 'Abdu, *Al-'Urwa al-Wuthqā*, 415.

41. See the introduction to the collected works of Jamāl al-Dīn al-Afghānī, *Al-A'māl al-Kāmila*, ed. Muḥammad 'Amāra (Cairo: Dār al-Kitāb al-'Arabī, 1968). Also see Keddie, *Islamic Response to Imperialism.*

42. See Jamāl al-Dīn al-Afghānī, *Khāṭirāt Jamāl al-Dīn al-Afghānī al-Ḥusaynī*, ed. Muḥammad Makhzūmī (Beirut: Dār al-Fikr al-Ḥadīth, 1965). This work by Al-Afghānī was first published in the 1930s.

43. For Al-Afghānī's critique of social Darwinism see Al-Afghānī, *Al-A'māl al-Kāmila*, 341 f.

44. Khān established the Aligarh Muslim University to promote his views and scientific education. One of his books, written in the wake of the mutiny of 1857, is on the merits of collaboration. See Aziz Ahmad, *Islamic Modernism in India and Pakistan, 1857–1964* (London: Oxford University Press, 1967); and Christian Troll, *Sayyid Aḥmad Khān: A Reinterpretation of Muslim Theology* (New Delhi: Vikas, 1978).

45. For example, *Al-Muqtaṭaf* published an article by a Lebanese religious scholar, Ḥusayn al-Jisr (d. 1909), who tried to understand Darwinism and reconcile it with the teachings of Islam. Ziadat, *Western Science in the Arab World*, 91–95.

46. For a commentary on the creationist movement and other contemporary Islamic views of science see Taner Edis, *An Illusion of Harmony: Science and Religion in Islam* (Amherst, NY: Prometheus Books, 2007), especially 115–63. Edis provides useful accounts of contemporary Turkish debates about creationism and, more broadly, about the relationship between Islam and science. However, his analysis of historical precedents is problematic, and he seems to be unaware of the substantial scholarship on the history of the exact Islamic sciences. For example, he maintains that Islamic science lacks theory and is no more than an "assemblage" of facts lacking explanation "within an overall theoretical scheme" (50). Perhaps the main weakness of Edis's analysis is his portrayal of contemporary cultic tendencies as normative trends in the relationship between science and Islam (107–8). For another approach that does not distinguish between contemporary and classical approaches see Pervz Hoodbhoy, *Islam and Science: Religious Orthodoxy and the Battle for Rationality* (London: Zed Books, 1991). Despite the assertion by many modern writers on Islam and science, most of the quasi-cultic contemporary approaches have no substantial historical precedents and are engendered by modernity and the social and political crises of contemporary Muslim societies. These cults have influence, but there are many more Islamic voices in

the modern period that still advocate the relative autonomy of science and religion.

47. Advocates of intelligent design argue that some forms of life are too complex to have simply evolved. Creationists also reject theories of evolution on the grounds that chance implies that life has no deliberate final meaning.

48. Yahya's other books include *The Evolutionary Deceit* and *The Disaster Darwinism Brought to Humanity.* A representative is Harun Yahya, *Why Darwinism Is Incompatible with the Qur'ān,* trans. Carl Rossini; ed. Jay Willoughby (Istanbul: Global, 2003).

49. The amount of material under Yahya's name includes hundreds of books, articles, DVDs, journals, and tapes. It is thus likely that this material was put together by a group of people and published under Yahya's name. In fact, firsthand accounts of Adnan Oktar suggest that he has a modest educational background, which seems to confirm that perhaps he did not write all of these materials. A fierce controversy erupted in Turkey between pro-evolutionists and members of BAV. The former asserted that Harun Yahya is a group of writers supported by BAV. This controversy is reported in Ümit Sayin and Aykut Kence, "Islamic Scientific Creationism: A New Challenge in Turkey," in *Reports of the National Center for Science Education* 19, no. 6 (1999): 18–29.

50. The group's spokesperson, Mustafa Akyol, publishes translations of works by American proponents of intelligent design and argues that believers in the East and in the West have intelligent design as a common cause. Yahya and his group have contacted American creationist organizations like the Institute for Creation Research in California to consult with them on strategies for getting creationism into school textbooks. The arguments, strategies, and rhetoric deployed by BAV are very similar to those deployed by American creationists. The only significant difference between the two is that Yahya is not wedded to the idea of a six-day creation, nor does he fix the history of the earth at six thousand or ten thousand years. The Qur'ānic account of creation is not as clear as the biblical one, hence this relative flexibility. Yahya also blames Darwinism for all evils in the world, including Nazism and terrorism. See Harun Yahya, *Facism: The Bloody Ideology of Darwinism* (Istanbul: Kültür Yayinlari, 2002); and Yahya, *Islam Denounces Terrorism* (Bristol, UK: Amal Press, 2002). In the last book Yahya blames Darwinist materialists for terrorism (147).

51. The causes for alarm over the spread of creationism in Europe was not due only to the widespread activities of BAV. In 2005, Vienna car-

dinal Christoph Schoenborn, a confidant of Pope Benedict, attacked neo-Darwinism, raising concerns that the Catholic Church may be aligning itself with intelligent design.

52. There are, to be sure, examples of more subtle and nuanced Islamic critiques of evolution. See, for example, Seyyed Hossein Nasr, *Knowledge and the Sacred* (Albany: State University of New York Press, 1987); and Osman Bakar, ed., *Critique of Evolutionary Theory: A Collection of Essays* (Kuala Lumpur: Islamic Academy of Science, 1987). See below for a discussion of the approaches of these scholars.

53. See, for example, 'Abd al-Razzāq Nawfal, *Al-Qur'ān wa al-'Ilm al-Ḥadīth* (Cairo: Dar al-Ma'arif, 1959), 24; and Dāwūd Sulaymān al-Sa'dī, *Athār al-Kawn fī al-Qur'ān* (Beirut: Dār al-Ḥarf al-'Arabī, 1999).

54. Bediuzzeman Said Nursi, *Sozler* (*The Words*) (Istanbul: Sinan Matbaasi, 1958).

55. Among Nursi's followers who continue to emphasize harmony between Islam and science is Fethullah Gülen, who has a relatively large following among Turkish Muslims inside and outside Turkey. See Fethullah Gulen, *Essentials of the Islamic Faith* (Fairfax, VA: Fountain, 2000).

56. Ṭantāwī Jawharī, *Al-Jawāhir fī Tafsīr al-Qur'ān al-Karīm*, 26 vols. (Cairo: Al-Ḥalabī, AH 1340–51), 2:483–84.

57. See Muzaffar Iqbal, "Islam and Modern Science: Questions at the Interface," in Ted Peters, Muzaffar Iqbal, and Syed Nomanul Haq, eds., *God, Life, and the Cosmos: Christian and Islamic Perspectives* (Burlington, VT: Ashgate, 2002), 3–41.

58. *Al-Sharq al-Awsat*, September 23, 2003. A more recent article in *The Muslim World League Journal* reports on the league's eighth International Conference on Scientific Signs in the Qur'ān and Sunnah, which was held in Kuwait. Ninety papers were presented, covering the following subjects: legislative and descriptive signs, human and social sciences, good tidings about the Prophet, geology and oceanography, astronomy, medical sciences, biology, and botany. The final statement issued by the organizers and participants of the conference made the following recommendations: to call upon scholars, researchers, and those interested in scientific signs in the world to adhere to the methodology of the International Commission on Scientific Signs in the Qur'ān and Sunnah; to encourage scholars of medicine and pharmacology to study "medication adopted by Prophet Muhammad"; to establish a satellite channel on scientific signs; to use scientific signs to proselytize; to organize special courses on this subject and "to urge universities and educational institutions to teach from the Book of Scientific Signs published by the Commis-

sion and to grant scholarships to the distinguished students who have special interest in this regard"; to organize an international contest every year; and to establish endowments for the support of activities related to this subject. The conveners concluded by thanking the emir of Kuwait and the king of Saudi Arabia for their generous financial support. *Muslim World League Journal* 35, no. 2 (2007): 11–13.

59. Similar assertions are made in relation to embryology, geology, astronomy, and other fields. In the mid-1990s, papers presented at a scientific miracles of the Qur'ān conference held in Pakistan covered such subjects as the use of jinn energy as an alternative source of power, and the development of a mathematical model to quantify the hypocrisy mentioned in the Qur'ān. A special interest to advocates of these views is parading the names of Western scientists who testify to the veracity of Qur'ānic statements on science and who often convert to Islam as a result of discovering in the Qur'ān the scientific facts that they have spent a lifetime studying. See, for example, Keith Moore et al., *Human Development as Described in the Qur'ān and Sunnah* (Mecca: Commission on Scientific Signs of the Qur'ān and Sunnah, 1992); Moore, *The Developing Human: Clinically Oriented Embryology. With Islamic Additions: Correlation Studies with Qur'ān and Ḥadīth, by Abdul Majeed Azzindani*, 3rd edition (Jeddah: Dar al-Qibla, 1983). Perhaps the most famous Western author who writes about the science of the Qur'ān is Maurice Bucaille; see Bucaille, *The Bible, the Qur'an and Science: The Holy Scriptures Examined in the Light of Modern Knowledge*, trans. Alastair D. Pannell and Maurice Bucaille (Indianapolis: American Trust Publications, 1979).

60. The main advocate of this exegetical approach as a way to guide future research is 'Abd al-Majid al-Zindani (Abdul Majeed Azzindani); see Al-Zindani, *Al-Mu'jiza al-'Ilmiyya fī al-Qur'ān wal-Sunna* (Cairo: Dar al-Madina, n.d.), 35. Right from the beginning, however, there were also many outspoken critics of the scientific exegesis school—Amīn al-Khūlī and many others. See, for example, J. J. G. Jansen, *The Interpretation of the Qur'ān in Modern Egypt* (Leiden: Brill, 1980), 53.

61. Tariq Suwaydān, *I'jāz al-Qur'ān al-Karīm: Min al-I'jāz al-'Adadī fī al-Qur'ān* (Beirut, n.d.).

62. Ziauddin Sardar, *Exploration in Islamic Science* (London: Mansell, 1989), 6. See also Ziauddin Sardar, *Islamic Futures* (London: Mansell, 1985).

63. Sardar, *Science and Technology in the Middle East*, 20–21. The terms and their translations are his.

64. Ibrahim Kalin, "Three Views of Science in the Islamic World," in Peters et al. *God, Life, and the Cosmos*, 43–75, 57–62. See also Leif Stenberg,

The Islamization of Science: Four Muslim Positions: Developing an Islamic Modernity, Lund Studies in History of Religions, no. 6 (New York: Coronet Books, 1996).

65. Sardar seems to look in the Qur'ān for an Arabic word that he thinks is relevant, and then assert that this word is a Qur'ānic principle, without making any attempt to justify his choice by reference to a tradition that cuts across religious, philosophical, and scientific genres. He reduces a rich tradition to his own arbitrary linguistic choices.

66. Kalin, "Three Views of Science," 63 ff. See also Osman Bakar, "Reformulating a Comprehensive Relationship between Religion and Science: An Islamic Perspective," *Islam and Science* 1, no. 1 (2003): 29–44. Bakar calls for revisiting "traditional Islamic disciplines such as epistemology, metaphysics, theology, cosmology, and psychology" to "formulate conceptual relationships between science and Islam" (39) and to "establish a philosophical framework for the harmonious relationship between the epistemological dimensions of science and the Islamic worldview as well as between ethical and societal dimensions of science and Shari'ah." Bakar identifies four components of science: "a body of knowledge, basic premises, methods of study and goals, all of which must be fully informed by the domain of imān and understood at the level of iḥsān" (29). Also see Osman Bakar, *Classification of Knowledge in Islam: A Study in Islamic Philosophies of Science* (Cambridge, UK: Islamic Texts Society, 1998); Syed Muhammad Naquib al-Attas, *Islam and the Philosophy of Science* (Kuala Lumpur: ISTAC, 1989); and Al-Attas, *Prolegomena to the Metaphysics of Islam: An Exposition of the Fundamental Elements of the Worldview of Islam* (Kuala Lumpur: ISTAC, 2001), chap. 3, 111–42. For an even more ambitious project see 'Adi Setia, "Islamic Science as a Scientific Research Program," *Islam and Science* 3, no. 1 (2005): 93–101; and Setia, "The Meaning of Islamic Science: Toward Operationalizing Islamization of Science," *Islam and Science* 5, no. 1 (2007): 23–52. In the last essay Setia calls for a "critical integration of the scientific endeavor into the conceptual framework of the Islamic worldview" and contends that "this programmatic redefinition of Islamic science will render it into a new over-arching 'paradigm' or 'research program'" (38–39).

67. Seyyed Hossein Nasr, "Islam and Modern Science," in Salem Azzam, ed., *Islam and Contemporary Society* (London: Longman, 1982), 179, 177–90. See also Nasr, *Islam and the Plight of Modern Man* (London: Longman, 1975); and Nasr, *The Need for Sacred Science* (Albany: State University of New York Press, 1993).

Index

'Abbāsids, 14, 45, 183*n*40
'Abdu, Muḥammad, 167, 169
active intellect, 102
'adadī, 172
Afghānī, Jamāl al-Dīn al-, 160–61, 165, 166–68
agricultural sciences, 47; agronomy, 47, 182–83*n*38
Al-As'ila wa'l-Ajwiba (Questions and Answers) (Al-Bīrūnī), 73–80
Al-Bashīr, 163, 165
algebra: arithmetization of, 40, 189*n*67; development of, 27–29, 39, 179*n*20; explained geometrically, 40–41; new beginnings in study of, 193*n*91
algebraic equations, theory of, 40
algebraic geometry, 51
Al-Ḥāwī fī al-Ṭibb (The Comprehensive Book on Medicine) (Al-Rāzī), 36–37
Aligarh Muslim University, 222*n*44
Al-Istidrāk 'alā Baṭlamyūs (Recapitulation Regarding Ptolemy), 68
Al-Jāmi' li-Mufradāt al-Adwiya wal-Aghdhiya (The Dictionary of Simple Medicines and Foods) (Ibn al-Bayṭār), 42–43
Al-Jawāhir fī Tafsīr al-Qur'ān al-Karīm (The Gems in the Interpretation of the Noble Qur'ān) (Jawharī), 170
Al-Jinān, 163
Al-Khalīl ibn Aḥmad al-Farāhīdī, 17
Al-Kulliyāt fī al-Ṭibb (Ibn Rushd), 93
Almagest (Ptolemy), 30, 32, 33, 58–59; critique of, 70; incorporating Hellenistic astronomy into one great synthesis, 61
Al-masā'il fī al-Ṭibb lil-Muta'allimīn (Questions on Medicine for Students), 30
Almohads, 197*n*24
Almoravids, 197*n*24
Al-Muqaddima (Ibn Khaldūn), 101–2, 143, 144
Al-Muqtaṭaf, 163–64, 165–66
Al-Qānūn al-Mas'ūdī (Al-Bīrūnī), 36
Al-Qānūn fī al-Ṭibb (The Canon of Medicine) (Ibn Sīnā), 37–38, 72

Al-Radd ʿalā al-Dahriyyīn (A Refutation of the Materialists), 166
Al-Riḥla al-Mashriqiyya (The Eastern Journey) (al-Ishbīlī), 43
Al-Shifā' (Ibn Sīnā), 71, 73
Al-Shukūk ʿalā Baṭlamyūs (Doubts on Ptolemy) (Ibn al-Haytham) 68, 71, 80, 86
Al-Zīj al-Dimashqī (Ḥabash), 33
Al-Zīj al-Ḥākimī al-Kabīr (Ibn Yunus), 35–36
Al-Zīj al-Mumtaḥan (The Verified Tables), 23
ancient sciences, sacredness and unity of knowledge in, 174
ʿaql (reason), 102–5
Arab Human Development Report (UNDP), 159
Arabic lexicon, development of, 17
Arabic science: continuity and coherence of traditions, 50–51; Persian and Indian influence on, 27, 29
Arabo-Islamic science: absence of, in the present, 150–51; beginnings of, 10–11, 13–15; bibliographical works of, 13; culture for, 15; history of, as part of larger, nonlinear history, 9–10; manuscripts of, mostly unavailable, 12; rise of, differing impressions of, 13–14; sudden appearance of, 15. *See also* Islamic science
Archimedes, 45
Aristotelian cosmology, 59–60, 62; critique of, 73, 75–76; questioning of, 83–84
Aristotelian cosmos, falling apart, 99–100
Aristotelianism, pure, 88–89
Aristotelian philosophy: critique of, 72, 75–77; reformulating, within demonstrative context, 88; Ṣadr arguing against, 137; used to critique adequacy of Ptolemy's planetary models, 70–71
Aristotle: on human reproduction, 90, 92; study of, 203–4*n*73; theories of, on vision, 39
Arithmetica (Diophantus), 28
Art of Algebra, The (Diophantus), 28
Ashʿarī school, 104, 143, 144
aṣl al-kabīra wal-ṣaghīra (large and small circles), 85
aṣl al-muḥīta (epicyclet), 85
astrolabes, 74
astrology, 114–15; criticism of, 114; Indian, 114–15; philosophical framework of, 130; political, 14; role of, in development of science, 14
astronomy, 32–36, 60–61, 65–66; Al-Birūnī's view of, prevailing, 80; alternative explanations, 134–35; autonomy of, 99, 137–38; Copernican, modern response to, 162–65; distinct from philosophy, 199*n*35; eastern reform tradition of, 64–67, 88, 99; first original Arabic work of, 29; Indian, 114–15; location of, among other sciences, 58; mathematical, 58–59, 84–85, 93; mathematical equivalence of models for, 81–82; multiple traditions feeding into study of, 29–30; natural, conceptual shift in treatment of, 84; not corresponding to reality, 87; as part of religious school curriculum, 22; practical, 13; practical disciplines in, 47–48; principles of, needing proof, 83; programmed study of, 22–26; reform of, having philosophical

foundations, 55; returning of, to its natural framework, 94–95, 99; right of, to derive principles from within its own discipline, 80; scholarship of, advances in, 32–36; in the service of Islam, 2–3; using, to study relationship of science and philosophy, 58–89; theoretical, 13, 93, 114; virtue of, deriving from nobility of its subject matter and from certainty of its proofs, 94–95; western reform tradition of, 64, 66–67, 88, 197*n*22
Atlas of Creation (Yahya), 168
atomism, doctrine of, 143
automatic machines, earliest book on, 45
Averroës. *See* Ibn Rushd
Averroësism, 204*n*73
azimuth, 177*n*3

Baghdādī, 'Abd al-Laṭīf al-, 186*n*51
Bakar, Osman, 226*n*66
Bannā, Ḥasan al-, 161–62
Bānū Mūsā brothers, 34, 45
being, discussion of, 136–37
Bilim Arastirma Vakfi (BAV), 168
biographical dictionaries, 20–21
Bīrūnī, Abū al-Rayḥān Muḥammad ibn Aḥmad al-, 36, 42, 49, 52, 67–68, 72–80, 93–94, 114–15, 138, 144, 147
Bitrūjī, Abū Isḥāq Nūr al-Dīn al-, 65, 66, 87, 99
blood, pulmonary circulation of, 179–80*n*21, 192*n*88
Book on the Solar Year, The, 33
botany, applied, 42–43
Brentjes, Sonja, 19
Bukhārī, Ṣadr al-Sharī'a al-, 68
burhān (demonstrative science), 89
Bustānī, Salīm al-, 163

calendar computations, 47–48
case medicine, 48
causality, attributed to habit and concomitance, 142–43
Center for Promotion of Scientific Research (ICPSR), 159–60
circular motion, uniformity of, as principle of natural philosophy, 95
civilization, engendering intellection, 108
classification, genre of, 100–101
clinical medicine, 48
colonialism, effect of, on Islamic science, 156–57
combinatorial analysis, 16–17
Committee on the Scientific Miracles of the Qur'ān and Sunna (Muslim World League), 171
Conrad, Lawrence, 13
contemplation, understanding of, 122
Copernican astronomy, counterrevolutionary nature of, 203*n*68
Copernicus, 65, 162–63, 164
Corruption (Aristotle), 79
cosmology, 59–60, 154
crafts, Greek attitude toward, 44
creation: beyond human comprehension, 122; continuous, 143; dependence of, on God, 125; marvels of, as recurrent theme in Qur'ānic commentaries, 121–22; presented in the Qur'ān, 119–21; Qur'ānic signs of, classification of, 122–23
creationist movement, in contemporary Islam, 168–69
crescent moon, visibility of, 33, 34, 47
cultural neutrality, ideal of, 147–48

Darwinism, Muslim response to, 165–69
De Caelo (On the Heavens) (Aristotle), 73, 75, 76
deferents, 61–64, 66
determinism, natural, denial of, 120
Dhanani, Alnoor, 129–30
diagnosis, medical, 48
Dioscorides, 43
divine philosophy, 107
dīwān, as bureaucratic term, 181–82*n*31
Dols, Michael, 13
doubts, genre of, 198–99*nn*31–32

earth: mobility of, 164; rotation of, 82, 84; seeking the center of the universe, 125; stationary position of, 59, 62–63, 83, 124–25
eccentrics, 61–64, 66, 85, 86–87
eclipse, prayers at, unrelated to natural phenomenon, 140
Elements (Euclid), 34
Endress, Gerhard, 13
epicycles, 61–64, 66, 85, 86–87
epicyclet, 85
epistemology, rethinking of, in light of development of knowledge, 106
Euclid, theories of, on vision, 39
exact sciences, role of, in Muslim education, 19
Exact Sciences in Antiquity, The (Neugebauer), 58
existence, discussion of, 136–37
experiential skill, engendering intellection, 108

Fārābī, Abū Naṣr al-, 89, 130
farā'id (algebra of inheritance), 20, 22
Farghānī, Al-, 32–33
Fāsī, 'Arabī Ibn 'Abd al-Salām al-, 5–8
Fikrī, 'Abd Allāh, 164
finitude, 124
Firdaws al-Ḥikma (Paradise of Wisdom) (Al-Ṭabarī), 31
first causes, 56, 100, 147
First Islamic Conference of the Ministers of Higher Education and Scientific Research, 159
First Philosophy, 103
foreign sciences, study of, 18

Galen: criticism of, 37; on human reproduction, 90–92
Generation (Aristotle), 79
geography, mathematical, 47
geometrical statics, Hellenistic studies on, influence of, 42
geometry: explained in algebraic terms, 40–41; knowledge of, not a legal obligation, 6
Ghāfiqī, Abū Ja'far al-, 42
Ghazālī, Abū Ḥāmid al-, 67, 104, 116, 117–18, 139–44, 147, 164
God: ability of, to disrupt natural order, 143; belief in, not contingent on a particular scientific view, 129; beyond human comprehension, 122; essence of, as subject of *kalām*, 136; oneness and omnipotence of, 119; winning the gifts of, 136; words used by, to refer to creation, 121
Greek philosophy, interest in, 113–14
Greek scientific tradition, influence of, on Arabo-Islamic sciences, 26, 27
Greek works: influence of, on Islamic thought and education, 18; introduction of, into Islam, 183–84*n*43
Gülen, Fethullah, 224*n*55
Gutas, Dimiti, 13–15

Ḥabash al-Ḥāsib, 33–34
ḥadīth: interest in, combined with medicine, 185*n*48; study of, as positive trait, 20
Handy Tables (Ptolemy), 29
hay'a (configuration), 114
heaven, as life form, inability to prove through rational proof, 141
heavenly objects, nature of, relevant to astronomy, 93
heavenly signs, discussions of, in Qur'ānic commentaries, 123
heavens: Ptolemy's mathematical models of, 59; unrelated to existence or production of scientific knowledge, 108
hereafter, knowledge of, 118
Hipparchus, 61
Hippocrates, on human reproduction, 90
hospital, as institutional achievement of Islamic society, 21
human intellect, 102–3
humans, benefits to: of creation, 120, 123–24; of heavenly signs, 123
Ḥunayn ibn Isḥāq, 30, 38
hydrodynamics, emergence of, 42
hydrostatics, 42

Ibn al-Bayṭār, 42–43
Ibn al-Haytham, 39, 67, 68–71, 72, 73, 88, 93
Ibn al-Khaṭīb, 143, 144
Ibn al-Shāṭir, 25–26, 65, 68
Ibn Bāja, 65–66, 86–87
Ibn Bannā' al-Marākishī, 154
Ibn Ḥazm, 116
Ibn Jamā'a, 'Izz al-Dīn, 185*n*48
Ibn Khaldūn, 101–2, 104–9, 143–48, 152–53
Ibn Maymūn (Maimonides), 86–87
Ibn Rushd (Averroës), 65–66, 87–89, 92–93, 131
Ibn Sīnā, 37–38, 67, 71–80, 89, 91–92, 102, 103, 130–31
Ibn Taymiyya, 104, 105–6, 116
Ibn Ṭufayl, 65–66, 87
Ibn Yunus, 35–36
ideas, new, received in cultural context, 11
Ihsanoglu, Ekmeleddin, 154–55
i'jāz (scientific miracle), 172
i'jāz raqamī, 172
Ījī, 'Aḍuḍ al-Dīn al-, 131–35, 164, 147
ijtihād (legal reasoning), 7–8
Ikhwān al-Ṣafā, 213*n*50
Ilkhānid Hulāgu, 24
Ilkhānid Zīj, 25
'ilm al-mīqāt (timekeeping), 20, 22
imported technologies, Muslim societies' reliance on, 158
Indian astrology/astronomy, 114–15
Indian science, 114
Indian tradition, influence of, on Arabic science, 27, 29
inductive reason, 106–7
Institute for Creation Research, 223*n*50
intellect: active, 102; human, 102–3; metaphysical, separate from human intellection, 104
intellection, types of, 206–7*n*108
intelligent design, 223*n*47, 224*n*51
International Conference on Scientific Signs in the Qur'ān and Sunnah, 224–25*n*58
international scholars, representing scientific tradition before period of translation, 15
Iqbāl, Muḥammad, 169–70

irrigation projects, 46–47
ISESCO, 159
Ishbīlī, Abū al-'Abbās al-, 43
Islam: culture of, science-philosophy relation in, 56; exact sciences in, study of, 12; intellectual authority in, 8–9; religious thought of, corresponding to conceptual scientific developments, 112; science-religion relation in, 110–11; scientific thought in, understanding the cultural significance of, 9
Islamic science: alternative narratives of, 13–14; culture of, related to other cultural forces, 10; decline of, 151–57; development of, 130, 112–13; legacy of, overlooked, 55; material support for development of, 14; momentum in, derived from interest in language sciences, 16–17; multiple scientific legacies feeding into, 27; practiced on unprecedented scale, 11–12; prepared for significant advances by early ninth century, 15–16; ten values of, 173–74; translation movement in, 10–11
Islamic theology: earliest interests in, 207*n*3; undermining metaphysical foundations of old sciences, 144
Isti'āb al-Wujūh al-Mumkina fī Ṣan'at al-Asṭurlāb (Comprehending All the Possible Aspects of the Craft of the Astrolabe) (Al-Bīrūnī), 74

Jābir Ibn Aflaḥ, 65–66
jadal (dialectics), 89
Jawharī, Ṭantāwī, 170
Jawziyya, Ibn Qayyim al-, 106
Jazarī, Al-, 46
Jibāra, Gabriel, 163–64
jiha (direction) of the *qibla*, 6–7
Jūzjānī, Abū 'Ubayd al-, 67

Ka'ba, 177*n*1
kalām, 117, 129–39; as apologetic exercise, 133–34; as incomplete philosophical system, 133–34; influence of astronomy on, 137–38; relation of, to other disciplines, 132–33; silence of, on natural matters, 134; subject of, as essence of God, 136. *See also* speculative theology
Karajī, Al-, 28, 40
Kennedy, Edward, 12
Khafrī, Shams al-Dīn al-, 67, 68, 80–82, 84, 85, 94, 144
Khān, Sayyid Aḥmad, 161, 166–68
Khaṭīb, Lisān al-Dīn Ibn al-, 152
Khāṭirāt Jamāl al-Dīn al-Afghānī (The Memoirs of Jamāl al-Dīn al-Afghānī), 167
Khayyām, Al-, 40, 41, 51
Khāzinī, Al-, 42
Khūnjī, Al-, 143–44
Khūrī, Naṣīr al-, 163
Khwārizmī, Muḥammad ibn Mūsā al-, 27–30, 39–40
Kindī, Abū Isḥāq al-, 38, 130
King, David, 2, 13
Kitāb al-'Ashr Maqālāt fī al-'Ayn (Ten Treatises on the Eye), 30
Kitāb al-Hay'a (Al-'Urḍī), 94
Kitāb al-Hya'a (Al-Bitrūjī), 99
Kitāb al-Jabr wal-Muqābala (Al-Khwārizmī), 27–28, 39–40
Kitāb al-Kāmil fī al-Ṣinā'a al-Ṭibbiyya (The Complete Book of Medical Art) (Al-Majūsī), 37

Kitāb al-Malakī (The Royal Book) (Al-Majūsī), 37
Kitāb al-Manāẓīr (Ibn al-Haytham), 39, 72
Kitāb al-Mawāqif fī 'Ilm al-Kalām (Al-Ijī), 131–32
Kitāb al-Taṣrīf li man 'Ajiza 'an al-Ta'līf (Al-Zahrāwī), 37
Kitāb fī al-Jadarī wal-Ḥaṣba (Al-Razi), 48
Kitāb fī Jawāmi' 'Ilm al-Nujūm (A Compendium of the Science of the Stars), 32
Kitāb fī Ma'rifat al-Ḥiyal al-Handasiyya (Al-Jazarī), 46
Kitāb Ṣuwar al-Kawākib al-Thābita (Al-Ṣūfī), 35
knowledge: considered a virtue, 113; constituting collective religious obligation, 113; hierarchy of, 89, 103; hybridization of, 27; incremental nature of, 108; organic unity of, no longer assumed, 99–100; theories of, 100–101; three systems of, 9; universal theoretical, 104; validity of form of, 107
knowledge base, formation of, 15, 16
knowledge systems, ranking of, no longer assumed, 147
Kuhn, Thomas, 52
kullī (knowledge), 103
kulliyāt (universal theoretical knowledge), 104

language, possible vs. real, 17
learning, disciplines of, relative to each other, 100–101
Lewis, Edmund, 165
life sciences, 13
Lives (Plutarch), 44
logic: acceptance of, 146; Islamic justification for, 143–44; possible errors in use of, 135–36
lunar model, alternative to Ptolemaic, 96–97
lunar visibility, 34

madrasas, devoted to legal sciences and other disciplines, 18
Māhānī, Al-, 40
Maḥmūd II, 155
Maimonides (Ibn Maymūn), 86–87
Majūsī, 'Alī ibn 'Abbās al-, 37
Makdisi, George, 18–19
Malik, 'Abd al-, 14
Mālik, Imām, 4, 7
Malikshāh, 24
Ma'mūn, Al-, 22–23
Manichaeans, 115
Manṣūr, Al-, 14, 29
manual labor, involving theoretical reasoning, 108
Maqālīd 'Ilm al-Hay'a (Al-Bīrūnī), 42
Marāghā school, 94, 196*n*18
Maṣmūdī, Al-, 3–4
mathematical astronomy, 32; distinct from philosophy, 74–75, 77, 78; focus on, 84–85; futility of harmonizing, with natural philosophy, 86; no clear link with natural science, 93; proceeding in disregard of physics, 98–99; self-evident principles of, 96
mathematical knowledge, precedence of, over religious authority, 2–3
mathematical methods: combining old and new, 49–50; inaccessibility of, to the masses, 4
mathematical principles, as tools for conceptualizing nature, 99

mathematical proofs, marginalizing natural philosophical arguments, 97
mathematical sciences: cross-application of, 34–35; practical disciplines in, 47–48
mathematics, 13; effects of cross-fertilization of disciplines, 40; as part of religious school curriculum, 22
Mawdūdī, Abū al-A'lā al-, 162
Mecca, determining the direction of, 1–2
mechanical devices, science of, 45–46
mechanical engineering, 45–46
mechanics, theoretical foundation for science of, 42
medical profession, prestige of, 21–22
medicine, 13; approaches to, indicative of adherence to Aristotelian philosophy, 90; early work in, 30–31; as practical science, 48; reform of, 203*n*72; scholarship of, advances in, 32–36; systematization of, 37–38; theoretical, 37–38
Menelaus theorem, 41, 49
mental judgments, abstract, validity of, 106–7
metaphysical reason, 102
metaphysics, 107; criticism of, 139; disengaging religion from, 140–41; disengaging science from, 144–45; occult sciences distinct from, 146; philosophical framework of, 130; reliant only on logic, 141; sovereignty of, over science, 175; unknown ultimate subject of, 145; unrelated to existence or production of scientific knowledge, 108
miracles, 143
modern science: intense responses to, 175–76; religious Islamic response to, 169–75
mosques: misalignment of, 2–3, 4–5, 7; orientation of, 3–5
motion, first mathematical proof of, 34
Muḥammad 'Alī Pāshā, 155–56
Muḥammad ibn Ibrāhim al-Fazārī, 29
mujtahid (scholar), 7–8
multiple possibilities, concept of, 144
Muslims: modern, calling for bringing science and religion together, 170; sense of security among, 27
Muslim world: destabilized by war, 152; economic decline in, 153–54; modernization of, 154–55, 156–57; plague affecting, 152–53; status of science in modern society, 157–60
Mu'tazilī school, 104, 183*n*40

Nabātī, Abū al-'Abbās al-, 42
Nallino, Carlo, 27
Nasr, Seyyid Hossein, 173, 174–75
natural bodies, containing principles of motion or stillness, 201*n*43
natural order, God's ability to disrupt, 143
natural phenomena, alternative explanations for, 126–28, 144
natural philosophy: futility of harmonizing, with mathematical astronomy, 86; rejection of, 81–82; related to practical discipline of medicine, 92–93; seeking proofs in, for astronomy, 83
natural principles, 97
natural sciences: presenting overall principles, 88; returning of astronomy to, 94–95; as root of all other sciences, 166
natural world, knowledge of, derived from the world itself, 143

nature: attempts to conform to, in mathematical astronomy, 98; laws of, not providing a moral code, 167; mathematization of, 57; as proof of the creator, 129
Neichiri sect, 161, 166–67
Neugebauer, Otto, 58
New Order, 155
new sciences, invention of, 39–43
normal science, 52
numerology, 172
Nursi, Bediuzzeman Said, 170

observatories, 22–26, 187–89*nn*60–62
occult sciences, 144–45. *See also* astrology; numerology
Oktar, Adnan, 168
optics, knowledge of, advances in, 38–39
orbs, planeted, 87
Ottoman empire, educational reforms in, 217*n*11
Ottomans, first alliance of, with Europeans, 155

patronage, as mode of supporting scientific activity, 23
Persian tradition, influence of, on Arabic science, 27, 29
pharmacology, developments in, 42–43
philosopher-peasants, 47
philosophers, mathematics education of, 131
philosophical theology, 132
philosophy: competing with religion, 130; conceptually separated from philosophy, 99–100; distinct from mathematical astronomy, 74–75, 77, 78; distinguishing it from mathematical sciences, 130–31; divine, 103; as inherited tradition and system of thought, 56; as professional practice or vocation, 56; relation of, with science, 54–55
Philosophy of Evolution and Progress, The (Shumayyil), 166
physical theory, 129–30
physics, 129–30; conflict of, with religion, 142; location of, in relation to other sciences, 57
Planetary Hypothesis, The (Ptolemy), 59, 70
planetary models: Al-ʻUrḍī's analysis of, 95; in conformity with natural principles, 97; studying, without linking to reality, 98
planetary motion, solid-body representations for, 98–99
planetary theory: in Ptolemaic model, 62–63; shift to, from practical astronomy, 65–66
planeted orbs, 87
planets, lacking influence on sublunar world, 128–29
Plutarch, 44
positivism, scientific: Islamic criticisms of, 173–75; understanding of science, 161–62
practical disciplines, examples of, 45
practical knowledge: epistemological rehabilitation of, 44–50; Greek attitude toward, 44
practical reason,105–6
practical sciences, Muslims' positive attitude toward, 44–45
principles: formulating, from within the discipline, 99–100; natural, 97; scientific, evolution of, 96
procedural reason, 102
prophets, 109, 146
prosneusis point, 63

Ptolemaic lunar model, alternative to, 96–97
Ptolemy, 29–30, 32, 58–59; Arab reexamination of, 32–33, 51; arguments of, distinguished from Al-Bīrūnī's, 79–80; astronomy of, Aristotle's cosmology underlying, 41, 59, 60, 62–63, 68–72; reassessing the astronomy of, 60–61, 63–72; theories of, on vision, 39

qibla, 147; determining the direction of, 1–8, 49; redirecting, without risking social conflict, 4–5; significance of debate over, 8–9
Qur'ān: attitude of, toward science, 115, 117; classical commentary on, 118–19; as comprehensive source of knowledge, 118; divine nature of, attempting to prove through modern science, 171; focus in, on physical universe, 170; four themes of, 119; interpreting allegorically, 168; lack of scientific interpretations of, 117, 118; linking, with science, 171–72; modern interpreters of, claiming it as source of scientific knowledge, 170–72; not used as source of knowledge about nature, 125–26, 127, 128–29; numerology of, 172; predicting modern scientific developments, 169, 170; providing no details on the natural order or ways to decipher it, 122; relation of, to modern science, 169–75; scientific miracles of, 169, 172; sign verses in, 123, 126–27; word order in, significance of, 120–21
Qur'ānic commentaries, 127–28
Qur'ānic exegesis, 116–29, 144
Qushjī, 'Alā' al-Dīn al-, 68, 82–83, 84, 94
Quṣṭā ibn Lūqā, 28, 38
Quṭb, Sayyid, 162

Ragep, Jamil, 82–83
rainbow, 88
Rashed, Roshdi, 13, 51
rational sciences: cultural and social spaces for practice of, 19; marginalization of, in Muslim societies, 18, 19; shared among all nations, 147; social prestige and respectability of, 20
Rāzī, Abū Bakr al-, 36–37, 48
Rāzī, Fakhr al-Dīn al-, 119–22, 125, 126–29
Rāzī, Muḥammad Ibn Zakariyyā al-, 79
realities, multiple, 81
reality, being reconciled with theory, 94
reason: centrality of, to theories of knowledge and classifications of science, 101; certain subjects outside jurisdiction of, 105; characteristics of, 104; distinctions in understanding of, 101; inductive, 106–7; lacking metaphysical existence, 108; limited scope of, 105; operative when applied to individualized material beings, 107; practical, 105–6; procedural approach to, 104; as tool, 104
religion: conflict of, with physics, 142; disengaging, from metaphysics, 140–41; as logical necessity, 109; scope of, including philosophical sciences, 139–40; vindicating, in the age of science, 173

religious authority, 3
religious debate, role of, 113–14
religious doctrine, effect of, on science, 114
religious knowledge, studied in its own context, 128
religious scholars, writing on science, 116
religious scholarship, combined with scientific scholarship, 19–20
religious sciences: classification of, 116; role of, in Muslim education, 19–20
religious writings, science addressed in, 116–17
Renan, Ernest, 160–61, 194*n*1
reproduction, theories on, 90–92
research and development, low investment in, in modern Muslim societies, 159–60

Sabra, A. I., 132
Ṣadr al-Sharī'a al-Bukhārī, 135–38, 185–86*n*50
Saint Joseph University (Beirut), 163, 165
Saliba, George, 12–15, 67, 81, 82, 94, 114
Samarqandī, Al-, 31
Sardar, Ziauddin, 173–74, 175
Ṣarrūf, Ya'qūb, 163
science: assessment of, by religious scholars, 116; claiming authority based on secular reason, 147; classifications of, 116; conceptually separated from philosophy, 99–100; cultural neutrality of, 147–48; cultural specificity of, 173–74; definition of, 56–57; disengaging it from metaphysics, 144–45; effect on, of religious doctrine, 114; epistemological requirements of, best described in texts, 12; European, 154–57; historical view of, as cultural artifact, 107–8; increasing professionalization of, 49; incremental nature of, 108; modern Islamic discourses on, displaying ignorance or abuse of history, 176; multiple interpretations and approaches to, awareness of, 31; partial, 103; perception of, in relation to religious knowledge, 114; philosophical classifications of, 103; positivist understanding of, 161–62; pursuit of, perceived as valuable in its own right, 21; as "rational sciences," 101; relation of, to philosophy, 54–55; religious position of, disputed, 117; status of, in modern Muslim societies, 157–60; sustaining cumulative growth in, 17; teaching of, venues for, 21; unity of, eroded, 109; universal principles of, 89
science-religion relationship, 110–14
Science Research Foundation, 168–69
science and technology indicators (STI) monitoring, 158
scientific activities, historical significance of, 11–12
scientific authority, 3
scientific discoveries, as proof of the creator, 129
scientific knowledge: communities of, 50–52; diffusion of, among educated elites, 49–50; divesting, of religious meaning, 146–47; existence and production of, unrelated to heavens or metaphysics, 108; expanding consumer base of, 49

scientific positivism, Islamic criticisms of, 173–75
scientific research, concurrent with translation movement, 26, 28–30
scientific scholarship, combined with religious scholarship, 19–20
scientific theory, function of, debates over, 55–56
scientists, liberated from trying to explore the unknowable, 109
sex organs, male and female, inverse similarity of, 90
Sezgin, Fuat, 13
shar' (law), 4
Shīrāzī, Quṭb al-Dīn al-, 68, 82–86, 98
Shumayyil, Shiblī, 165
sines, theorem of, discovered simultaneously by three astronomers, 52
souls, separate existence of, 142
speculative theology: disengaging specialized sciences from philosophical framework, 138; as genuine form of knowledge, 132; replacing Aristotelian philosophy, 131. See also *kalām*
spiritual essences, outside the realm of knowledge systems, 146
statics, dynamic approach to study of, 42
subjugation, 119–20, 123–24
Ṣūfī, 'Abd al-Raḥmān al-, 35
Sufis, influence of, on cosmology, 154
sun, motion of, in Ptolemaic model, 62
surveying, 46–47
Suyūṭī, Jalāl al-Dīn al-, 117–18
Syriac, speakers of, important role of, as translators, 14–15
Syrian Protestant College (Beirut), 163, 165
systematic mathematization, 35
Ṭabarī, 'Alī ibn Sahl Rabbān al-, 31
Ta'dīl al-'Ulūm (The Adjustment of the Sciences (Ṣadr), 138
tafsīr (Qur'ānic exegesis), 116–29, 144
Tahāfut al-Falāsifa (The Incoherence of the Philosophers) (Al-Ghazālī), 139
Tājūrī, Al-, 5–6
Tarkīb al-Aflāk (Al-Jūzjānī), 67
technical faculties, engendering intellection, 108
Thābit ibn Qurra, 33, 34, 38, 40
theological debate, role of, 113–14
theology: increasingly incorporating philosophy and astronomy, 131; interaction of, with other disciplines, 132–33; not intended as complete philosophical system, 137; philosophical, 132; scope of, redefined and transformed, 138–39; speculative (see *kalām;* speculative theology); works of, gaining in complexity and systematization, 131
theoretical knowledge, higher than practical knowledge, 103
theoretical medicine, 37–38
timekeeping, 25–26, 48, 192*n*84
translation movement: aspect of emerging scientific culture, 14–15; causing momentum in development of Islamic sciences, 16–17; as complex phenomenon, 14; concurrent with Islamic scientific research, 26, 28–30; in Islamic sciences, 10–11; methods used in, 16; results of, 16; specialized lexicons developed during, 16
transmitted evidence, as starting point for *kalām*, 133
transmitted knowledge, 136

trigonometry: advances in, essential to development of Arabic astronomy, 36; Arab scientists' enrichment of, 41; developments in, 35–36; emergence of, as independent science, 41–42
Tuḥfa (Al-Shīrāzī), 85
Ṭūsī, Naṣīr al-Dīn al-, 24, 42, 63, 68, 83
Ṭūsī, Sharaf al-Din al-, 40, 51
Ṭūsī couple, 65, 85, 98

Ullmann, M., 13
Umayyad administration, Arabization of, 14, 15
United Arab Republic Charter, 22
universal principles, deriving from abstracted metaphysical intellect, not applicable to natural knowledge, 107
universals: as mental constructs without real existence, 106; science of, 103
universal science, partial sciences dependent upon, 103
universe, organic, 87–88
'Urḍī, Mu'ayyad al-Din al-, 65, 68, 94–98
uṣūl (principles), of astronomy, 95–96

vacuum, existence of, 200*n*40
vision, Hellenistic theories of, rejected, 39

weights, science of, 42, 191*n*77
Weinberg, Steven, 214–15*n*2
Western science, acceptance of, in modern era, 162
worship, mathematical problems related to, 47–48. See also *qibla*

Yahya, Harun, 168
Ya'qūb ibn Ṭāriq, 29
Yūḥannā ibn Māsawayh, 30, 38

Zahrāwī, Abū al-Qāsim al-, 37
Zarqālī plate, 178*n*18
zenith, 177*n*2
Zīj al-Shāh, 29
Zīj al-Sindhind, 29
Zīj al-Sindhind (Al-Khwārizmī), 29